Lecture Notes in Computer Science 564

Edited by G. Goos and J. Hartmanis

Advisory Board: W. Brauer   D. Gries   J. Stoer

I. Herman

The Use of
Projective Geometry
in Computer Graphics

Springer-Verlag

Berlin Heidelberg New York
London Paris Tokyo
Hong Kong Barcelona
Budapest

Series Editors

Gerhard Goos
Universität Karlsruhe
Postfach 69 80
Vincenz-Priessnitz-Straße 1
W-7500 Karlsruhe, FRG

Juris Hartmanis
Department of Computer Science
Cornell University
5148 Upson Hall
Ithaca, NY 14853, USA

Author

Ivan Herman
Centre of Mathematics and Computer Science
Kruislaan 413, 1098 SJ Amsterdam, The Netherlands

CR Subject Classification (1991): I.3.2-4, G.0, I.6-7

ISBN 3-540-55075-5 Springer-Verlag Berlin Heidelberg New York
ISBN 0-387-55075-5 Springer-Verlag New York Berlin Heidelberg

Typesetting: Camera ready by author
Printing and binding: Druckhaus Beltz, Hemsbach/Bergstr.
45/3140-543210 - Printed on acid-free paper

Preface

This book has its history. In 1986, I got involved with some of my friends in the implementation of a GKS-3D package at a small firm called Insotec Consult, in Munich, Germany. We did not have too much time, so we tried to produce a straightforward implementation of the Standard as fast as we could and we planned to optimize it in a later stage. We already had the common experience of a GKS-2D implementation, so the task seemed to be relatively easy. To implement the 3D output pipeline, we just had to program some appropriate matrix-matrix and matrix-vector multiplications and we soon started the first toy test-program, to navigate among randomly positioned wire-framed objects in space. The output seemed to be right at a first glance: we saw what we expected. However, in some cases some unexpected lines appeared on the screen in a fairly chaotic manner which we just could not explain to ourselves. Being self-critical enough, we set down to find the bug in our code by running through it several times, but without success. It was only after several days' of work that we begun to suspect that implementing a 3D pipeline is not that simple after all.

Of course, we did not do our homework properly; the notion of 'external lines' or, as it is called in this book, the 'W-Wraparound' (which turned out to be the reason for our problems) had already been presented in the literature before. However, not having easy access to literature in our small software house we had to find a solution for ourselves. This work had unexpected results for all of us; it made us realize that the computer graphics community makes astonishingly little use of the notions of projective geometry in solving its problems in spite of the fact that the very essence of 3D computer graphics is closely related to this classical branch of mathematics. Looking at the problem with fresh eyes has proven to be advantageous after all; it has led us to some really new and interesting results which have proven to be of a general interest, going beyond the particular problem which had triggered it. This story was was the start for me of a long-term activity in trying to adapt some projective geometry results for the purposes of computer graphics; it led to a series of publications and to my PhD thesis which I defended in 1990 at the University of Leiden, in the Netherlands. This thesis has essentially provided the material for this book.

Such a small book obviously cannot and does not cover all the problems of a 3D output pipeline as it appears in practice, nor does it describe all possible applications of projective geometry in computer graphics. Instead, it concentrates on some of its special and very much algorithmic aspects which are related first of all to the different transformations used in computer graphics. However, all specialists or students of computer graphics who want to understand the underlying mathematic principles of 3D graphics systems or want to participate in the implementation of a new one can get, I hope, some inspiration for such work.

A hidden, but also very important aim of the book is to make clear what the usual curricula in computer sciences seem not to emphasize enough: that higher-level mathematics — projective geometry is just an example — have a major role to play in computer graphics and computer science in general. A number of problems become easier to solve or just simply to describe provided that the appropriate mathematical tools are used. If this book succeeds in turning the attention of some of its readers toward mathematics again, it has achieved a major goal set for myself when planning to publish this work through Springer Verlag.

The mistake of the 'missing homework' was committed together with my colleague and friend, J. Reviczky, with whom I had a long and very fruitful cooperation before I decided to join the Center for Mathematics and Computer Sciences in Amsterdam; it is a pleasure for me to acknowledge his major role in the birth of this book. The ongoing discussions and common works with all my former colleagues of Insotec Consult, primarily J. Hübl, are also acknowledged and very much appreciated.

The text of this book was fleshed out, revised and shaped in discussions with my PhD advisers, namely Prof. F. Peters, Prof. R. Hubbold, Prof. G. Joubert, Prof. D. Duce and Prof. J. van den Bos. It is again a pleasure for me to acknowledge their help and encouraging remarks in the preparation of the manuscript. The outstanding facilities of the Center for Mathematics and Computer Sciences made it possible to produce the output reliable and the way I wanted. Finally, I am grateful to my wife, Eva, who really pushed me to do this work and helped me through difficult times; this work would never have been finished without her.

Amsterdam, October 1991, Ivan Herman

Contents

1. Introduction

The ultimate goal of all 3D graphics systems is to render 3D objects on an inherently two dimensional surface, which may be a plotter output, a screen or anything similar. One way of achieving this is to start from the medium itself, and to cast so–called *rays* (that is half–lines), starting from the viewer's eye and determined by each raster point of the display, back into the scene. The intersections of these rays with the three dimensional objects will determine the visible raster dots on the view surface. This is basically what the technique called *ray–tracing* does. Although ray–tracing gives the possibility to perform complicated shading, transparency, translucency, and shadow calculations on all intersection points and can therefore lead to highly realistic images, the computational requirements of this approach are still so high that it cannot be used for interactive applications (for a more consistent description of ray–tracing see for example [Fole90] or some other standard textbooks on computer graphics).

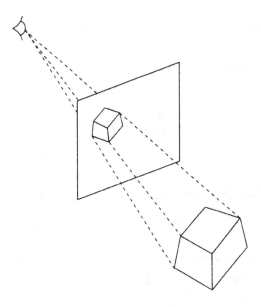

Figure 1.1.

Another approach, which is adopted by most 3D graphics systems as well as the different ISO documents on 3D graphics (for example GKS–3D, PHIGS, PHIGS PLUS, CGM Addenda etc; additionally to the ISO documents themselves, the reader may refer for example to [Arno90] for a good overview of these standards) is to design a system which performs a central or a parallel projection of the objects to render onto the view surface. As this projection is done on the geometrical level (that is by describing the projected lines/polygons etc. on the 2D surface), this approach requires fewer calculations than ray–tracing. As a consequence however, these three dimensional systems have to make use of the mathematical results of *projective geometry*.

Early three dimensional systems explicitly implemented the parallel or central projections as shown in figure 1.1, that is they calculated the intersection points of the projection lines with the view surface to determine the outlines of the projected objects. In more modern systems as well as in all the cited ISO documents the approach is different. These systems define and perform a transformation of the whole scene, so that the central projection becomes a simple parallel one onto the $z=0$ plane. More precisely, the transformation should result in a projection such that:

$$(x,y,z)^T \rightarrow (x,y)^T \tag{1.1}$$

This means, mathematically, that the transformation to be applied is such that the view point is moved to the "point at infinity on the z–axis". The reason for this approach is that performing the parallel projection after this transformation creates the possibility to perform the so called Hidden Surface/Hidden Line calculations efficiently, that is to determine which part of the scene is effectively visible on the view surface (this presupposes, however, that the algorithms make use of the relative magnitude of the z values only). It also makes the necessary clipping processes computationally simpler. Furthermore, efficient hardware can also be used to perform these computations (for example the so–called *Z–buffer* calculations) easily and effectively (for the details, see the cited ISO documents in [ISO88], [ISO88b], [ISO89], [ISO89a] or any of the general works on computer graphics like [Newm79], [Fole84], [Magn86], [Mudu86], [Salm87], [Watt89], [Fole90]).

The mathematically precise formulation of what is stated above is to introduce the notion of ideal points, which are the mathematically precise definitions of the "points at infinity". *Projective geometry* is the branch of mathematics which provides a uniform description of ideal and Euclidean points. One may also speak of projective planes (ie Euclidean planes extended with ideal points) and projective spaces (ie Euclidean space extended with ideal points).

Using these notions one can say that all 3D graphics systems tend to perform their internal calculations in the projective space rather than the Euclidean one. The fact that a projective space has its natural coordinate system by means of homogeneous coordinates makes this approach feasible: if a homogeneous coordinate system is chosen, all projective transformations may be described by (homogeneous) 4x4 matrices and the effect of the transformation is then a matrix–vector multiplication. The use of homogeneous coordinates also gives a uniform way to describe efficiently all transformations usually used in graphics systems: rotations, scaling, shearing and translations.

Although a number of calculations (ie line/line intersections and the like) can be performed easily with homogeneous coordinates, the fact of working in projective rather than Euclidean space is the source of additional complications. A number of graphical output primitives are described inherently in a linear fashion, that is using different linear combinations of the points and vectors involved in 3D. Examples are the description of colour patterns (usually a rectangular array is to be calculated) and high precision text (described as a set of small line segments). While a "traditional" transformation of a Euclidean space does keep this linear structure, this is definitely not the case for projective space and projective transformations, which might lead to distortions which are not easily describable by linear means. As a consequence, the implementor of a graphics system might choose to perform all such generation (that is to generate the series of points approximating the high precision text) before the transformation and to transform, as a second step, the generated points. This leads clearly to a loss of efficiency: the number of generated points may be very high and therefore the computational cost of the transformation itself may be significant.

The use of projective transformations and projective space may lead to other problems. In the course of a projective transformation it may happen that some of the (originally Euclidean) points of the objects to be rendered become ideal points (in homogeneous terms this means that the last coordinate value, usually denoted by w, will become 0). A graphics system must have some means to deal with such situations which can lead, if not properly handled, to computational singularities and unexpected visual effects on the screen.

A common framework to handle these problems may be found if the exact mathematical behaviour of projective transformations in graphics systems is carefully analysed. The derivation of such a framework is the central subject of this thesis. It will be shown that a mathematically precise description of the projective geometrical nature of a 3D (or even 2D) graphics system leads not only to a deeper understanding of the system but also to new approaches which result in faster or more precise algorithms.

The thesis aims to be self–consistent for all those computer scientists who have a general knowledge about computer graphics. However, no preliminary knowledge about projective geometry is needed; instead, an extensive introduction to the subject with all necessary notions and theorems will be given. It is well–known that projective geometry is (unfortunately) missing from the usual curriculum for computer scientists and apart from the excellent book of Penna and Patterson ([Penn86]) and the tutorial at the Eurographics'88 Conference ([Herm91]) there are no other textbooks available for specialists of computer graphics. It is of course true that a large number of textbooks are available as an introduction to projective geometry; some of them are listed among the references ([Coxe49], [Coxe74], [Crem60], [Fisc85], [Hajó60], [Heyt63], [Kell63], [Keré66], [Lanc70], [Rose63], [Stru53]) and the list is certainly not exhaustive. (The author's knowledge of this area originates mostly from the lectures made by Prof. M. Bognár at the University of Budapest in the year 1971, from which only hand–written notes are available.) A common problem with these books is that they were written *by* mathematicians *for* mathematicians; in other words they stress different aspects of projective geometry from those needed for the purposes of computer graphics. Whereas they present very general and important theorems, the *computational* aspects of the mathematical structures tend to be disregarded; in other words, it is sometimes quite difficult to extract the specific information which is necessary for the purposes of computer graphics. Those readers, however, who have the necessary background in projective geometry, may bypass chapter 2 altogether and start at chapter 3 directly.

The material included in this thesis has been admittedly influenced by the algorithmic problems arising when implementing one of the ISO 3D graphics standards. Some of the ideas have been implemented when the author took part in a GKS-3D and a CGI-3D implementation in the years 1986–87 at Insotec Consult GmbH in Munich (Federal Republic of Germany) (see also [Herm88b]). PHIGS and PHIGS PLUS have also generated a number of additional problems while some of the works presented here (and having been published already elsewhere, like [Herm89]). However, most of the problems presented in the sequel are not exclusively proprietary to the standardisation activities; in fact, they are inherent to most 3D graphics systems in use and/or in development such as the Doré package of Ardent Computers [Arde87], RenderMan of PIXAR [Pixa88] and others still to come[†].

[†]The main outlines of all the cited ISO standards are however considered as known and will not be detailed in what follows. The reader should refer to the relevant ISO documents or the existing textbooks and tutorials on the subject, as for example [Hopg83], [Salm87], [Arno90], [Hubb90], [Howa91] or [Hopg91].

2. Projective Geometry in General

2.1. Some History

Projective geometry is by no means a new field of mathematics. Some of its very classic theorems cited in all basic textbooks (eg theorems of Pappos and Menelaos) are of Greek origin. It can be supposed that ancient Greek mathematicians knew much more about projective geometry; unfortunately, most of their work has been lost and only some indirect facts can be used to measure the exact amount of their knowledge.

It was in the period of the Renaissance that projective geometry gained a much greater importance. This was the result of the fact that artists at that time were greatly interested in creating realistic pictures; as opposed to medieval painters they wanted to understand exactly how a three dimensional object could be rendered on a two dimensional plane, that is a canvas. And this is basically what projective geometry is all about. At a time when it was (still) natural that artists would also work on "scientific problems" if they felt the necessity for it, Pietro della Francesca (1420-1492) or Albrecht Dürer (1471-1528) wrote down their ideas about the rules of projection; the book of Dürer ([Düre66]) is probably one of the first books ever written on the subject. It is also interesting to note that he produced a number of carvings in which practical techniques of how to make projections are presented in an artistic way.

A more concise mathematical investigation on projective geometry was started by G. Desargues (1593-1662). It was he who has introduced the notion of a point and a line at infinity. His work was followed by a number of other mathematicians (B. Pascal, L.N. Carnot, G. Monge and others) who discovered those theorems and facts about projective geometry which still form the basis of the theory today.

However, the exact role of projective geometry in the description of the surrounding world was for a long time somewhat fuzzy. Indeed, one has to understand that at that time geometry as well as the philosophy of nature in general was very much dominated by Euclidean geometry. In his monumental work *The Elements* ([Eukl75]), which was a kind of an encyclopaedia of Greek geometry, Euclid had created the first axiomatic system in history of mathematics; his work was so successful that up to the 19th century everybody thought that Euclidean geometry was not only an efficient tool to describe nature but the surrounding world *was* Euclidean. This belief was even more strengthened by the fact that Newton's *Principia* was very much based on Euclidean geometry when describing the rules of mechanics. Very typically, the "Princeps Mathematicae", F. Gauss (1777-1855), who was probably one of the first mathematicians to realise that the truth might be different, did not dare to publish his results; he was afraid to be in conflict with all the great intellects of his time.

This firm belief in the overall nature of Euclidean geometry was tarnished by the independent works of J. Bólyai and N.I. Lobatchewsky. Indeed, these two mathematicians succeeded in creating a new geometry (which later received the name of *hyperbolic geometry*) which was fundamentally different from the Euclidean one. This geometry was the result of their investigations on Euclid's so called 5th postulate, which said that if a point does not intersect a line, then there is *only one* line intersecting this point which is parallel to the given line. After centuries of unsuccessful attempts to *prove* this postulate out of the remaining axioms of Euclidean geometry, both Bólyai and Lobatchewsky created a new axiomatic system by taking all other Euclidean axioms and the *negation* of the 5th postulate (that is that there exist *more than one* distinct parallel line intersecting the external point). This new geometry is very different in flavour: the sum of the internal angles of a triangle is not 180°, the "traditional" trigonometrical equations are no longer valid etc. It is, however, undecidable whether Euclidean geometry or the hyperbolic one is the adequate description of reality; in fact, both of them are only *models* and one could as well describe the whole surrounding world by using hyperbolic geometry instead of the Euclidean one.

Besides the technical nature of this result, the birth of hyperbolic geometry has shown that there might be a whole range of different "geometries", each of them modelling some particular aspect of reality. One may speak of multidimensional geometry (giving a geometrical structure to the set of vectors of higher dimensions), of complex geometry, of hyperbolic and elliptic geometries etc. Projective geometry turned out to be one of these different geometries, a useful tool in the description of some phenomena concerning projections, lines, planes and so forth.

These results (together with other advances in mathematics in the 19th century) had also strengthened the need of a precise foundation of all mathematical fields, including projective geometry. This was performed by the thorough use of the axiomatic method and of set theory, which has become the fundamental basis of mathematics in our century and had been initiated largely by the *Erlanger Programm* of F. Klein (1849–1925) and by the *Grundlagen der Geometrie* of D. Hilbert (1862–1943). Projective geometry has also been reformulated in this way: there is a very precise set of axioms which defines projective geometry and the existence of this axiomatic approach gives also a very precise insight of how projective geometry is related to other fields of mathematics (specially Euclidean geometry). This system of axioms will be presented in a later section.

In the 20th century traditional projective geometry has lost its momentum as a field of basic mathematical research; in this sense it might be considered as a "classical" theory[†]. It has by no means lost importance though, having given birth to a whole range of practical tools and methods used to make drafts and technical drawings throughout the world. It is the very aim of this present work to show that a more precise knowledge of projective geometry can also play a significant role in

---

[†] Today's researches are more directed toward algebraic and combinatorical problems arising when investigating for example *finite* projective spaces; these problems have had, however, no relevance for computer graphics up to now.

giving new and perhaps more understandable approaches to the methods and algorithms used in computer graphics.

2.2. Notational Conventions

Before going further some notations are listed which will be used throughout the thesis.

Points will be denoted by capital Latin characters (A, A', A'', P, Q and also A_i, B_j etc.) whereas small Latin characters (a, a', a'', n, l etc.) will denote lines. 2D subplanes of the Euclidean space will be denoted by capital Greek letters like Π, Π', Π'', Φ, Ψ. The symbol "$\wedge$" will be used to denote intersection; that is $a \wedge b$ will give the intersection of the lines a and b while $\Pi \wedge b$ will denote the intersection of the plane Π and the line b. Symmetrically, the symbol "$\vee$" will be used for a generated line or plane; that is, $P \vee Q$ will give the line generated by the points P and Q while $R \vee a$ denotes the plane generated by the point R and the line a. The symbols $\wedge$ and $\vee$ will also be used in logical statements; for example $A \wedge a = \emptyset$ means that the intersection of the point A and the line a is the empty set, that is the point A does not belong to the line a. Both the relation "$\wedge$" and "$\vee$" are associative and, consequently, their meaning can be extended to more than two operands. This means for example that the notation $P \vee Q \vee S$ can be used to denote the plane generated by the points P,Q and S. Finally, $I\!E^2$ will be used to denote the 2D Euclidean plane in general whereas $I\!E^3$ will be the Euclidean (three dimensional) space.

The set of real numbers will be denoted by $I\!R$ and the symbols $I\!R^2$, $I\!R^3$, $I\!R^n$ etc. will denote the set of *column* vectors of corresponding dimensions. The vectors themselves will be denoted by small Latin characters as well, choosing characters usually at the end of the alphabet. If $x \in I\!R^3$, x^T denotes the *transpose* of the vector x. To save space, $(1,2,3)^T$ will be used instead of

$$\begin{pmatrix} 1 \\ 2 \\ 3 \end{pmatrix} \tag{2.1}$$

To resolve ambiguities, the notation $\vec{x}$ will also be used for a vector to distinguish it from a line. In a number of cases a coordinate system will be implicitly present in the geometric environment in use. In such cases, the points will be identified with their coordinate vector and characters like p or q, which denote in fact vectors, will also be used to denote points. This convention, although not necessarily very precise mathematically, will be extensively used later.

Matrices will be denoted by capital Latin characters with their elements being the corresponding small characters (that is the elements of the matrix A will be $a_{i,j}$). A^T denotes the transposed matrix of A; if x is a vector of the appropriate dimension, Ax will be the (multiplied) column vector and $x^T A$ the row vector. In case of doubt, the notation $\bar{A}$ will also be used to differentiate matrices from points.

The scalar (or inner) product of two vectors x and y will be denoted by $x^T y$; the vector (or outer) product of two vectors is $x \times y$. This latter product can be calculated by evaluating the following formal determinant:

$$x \times y = \det \begin{bmatrix} x_1 & x_2 & x_3 \\ y_1 & y_2 & y_3 \\ e_1 & e_2 & e_3 \end{bmatrix} \tag{2.2}$$

where e_i, $(i = 1,2,3)$ denote the basic unit vectors of $\mathbb{R}^3$ (that is $(1,0,0)^T$, $(0,1,0)^T$ and $(0,0,1)^T$). In other words, the vector product is:

$$x \times y = \left[\det \begin{bmatrix} x_2 & x_3 \\ y_2 & y_3 \end{bmatrix}, -\det \begin{bmatrix} x_1 & x_3 \\ y_1 & y_3 \end{bmatrix}, \det \begin{bmatrix} x_1 & x_2 \\ y_1 & y_2 \end{bmatrix} \right]^T \tag{2.3}$$

If A is a quadratic matrix of dimension n and $x,y \in \mathbb{R}^n$, xAy will denote the so called *bilinear form*, that is:

$$xAy = x^T(Ay) = (x^T A)y \tag{2.4}$$

2.3. The Axiomatic System of Projective Geometry

2.3.1. Background

Figure 2.1 illustrates what were the basic problems which led to the development of projective geometry. A central projection is made onto the plane Π; for the sake of simplicity we concentrate now on projecting (from the centre C) the plane Ψ onto Π.

The central projection has a number of very nice properties as seen from the figure. It maps (almost) all points of Ψ onto points of Π; it maps a line of Ψ onto a line of Π and maps (usually) the points of intersections of two lines onto the point of intersection of the image of these lines. It is also almost invertible; that is for almost all points of Π there is a corresponding point of Ψ which would be the inverse image.

Figure 2.1 also shows why such vague statements are to be used to characterise this mapping. Indeed, there are some points of Ψ (eg P) for which the central projection is not properly defined (the projection line does not have an intersection with the image plane). Accordingly, all lines which intersect in points for which no image could be defined (like m and n), though being mapped onto lines, become parallel on Π, that is they have no intersection points any more. Additionally, all points on the line l' on Π (which is the intersection line of Π and a plane containing C and parallel to Ψ) are without inverse image.

The reason for all these singularities can be traced back to the very existence of parallel lines in a Euclidean environment. One would like to interpret somehow what happens to the intersection point of parallel lines; if this were done, the image of P could be defined as being the "intersection" in some sense of the parallel lines of m' and n'. Clearly, the problem is that no Euclidean point can play such a role; there are "holes", or missing elements in the set of all points in a Euclidean plane.

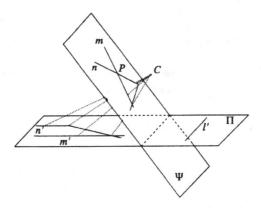

Figure 2.1.

What is usually done in such cases in mathematical practice is to *enlarge* the basic set one is working with. In other words, a new set is created which would contain (in this case) the set of all Euclidean planar points but which would also contain some additional elements. The usual notions (in this case "lines", "intersection" etc.) should be *extended* for this larger set so as to include the traditional notions as well. If this extension is well done, the "holes" may be filled and one arrives at a much clearer structure than the original one. This is what will be done for projective geometry: new elements (which are not Euclidean points) will be defined and the notion of lines, line intersections etc. will be extended so as to include these new elements as well, such that all the problems stated in figure 2.1 can be overcome. These new elements will be called the *ideal points*; they will be the mathematically precise form of what is commonly called "points at infinity".

The way of doing this extension is again very classic in mathematics but it requires some abstractions which are not always easy to understand for non−mathematicians. What has to be used is what is called an equivalence relation and the generated quotient set. This is as follows.

A (binary) *relation* on a set H is defined to be a subset of the set of element pairs (that is a subset of $H{\times}H$). If $x,y{\in}H$ and the relation is denoted by ρ, than $x\rho y$ denotes the fact that the two elements x and y belong to the same subset, that is the defined relation "holds" for them. A relation ρ is an *equivalence relation* if the following three properties hold:

$$\forall x,y,z : \ x\rho y \wedge y\rho z \Rightarrow x\rho z$$
$$\forall x,y \ \ : \ x\rho y \Rightarrow y\rho x \qquad\qquad (2.5)$$
$$\forall x \ \ \ \ \ : \ x\rho x$$

The relation is said to be transitive, symmetric and reflexive. One can easily recognise that the relation "=" on $\mathbb{R}$ is an equivalence relation.

Equivalence relations divide the set on which they are defined into a set of mutually disjoint subsets (they provide an abstract tessellation of the supporting set). Indeed, if A is an arbitrary set and ρ is a relation defined on A for which (2.5) holds, then for each $x \in A$ the following set can be defined:

$$x_\rho = \{ z : z \in A \text{ and } x\rho z \} \tag{2.6}$$

These sets are called the *equivalence classes* of A defined by ρ. It is a typical mathematical exercise to prove that if $x\rho y$ holds then also $x_\rho = y_\rho$ and if $x\rho y$ does not hold then $x_\rho \cap y_\rho = \emptyset$. In other words, the equivalence classes are disjoint sets which are "generated" by each elements of the set A. As a result of the reflexivity, $x \in x_\rho$; that is, the equivalence classes effectively tessellate the whole set (no element is left out). Proving the previous statements is not particularly complicated; this is left to the reader.

As a result of the tessellation one can also speak of a new set, denoted by A_ρ, by taking:

$$A_\rho = \{x_\rho : x \in A\} \tag{2.7}$$

This set is usually called the *quotient set* or *quotient space*. If one wants to go beyond the abstract definition, it could be said that it is a set of which the elements characterise the equivalence relation ρ by collecting into one element all elements of A which somehow belong together.

The quotient set is widely used for a mathematically precise formulation of an extension mechanism. An equivalence relation is defined either on the elements of the set A or on some other set B related strongly to A; the set $A \cup B_\rho$ provides then an extension of A which makes the characterisation of the relation ρ simpler (provided the set B is chosen in an appropriate manner). This approach is very widespread in mathematics; this is how for example irrational numbers are constructed out of rational ones on $\mathbb{R}$, and this is also how ideal points are defined properly.

2.3.2. The Basic Construction

For the sake of simplicity, in what follows the axiomatic system for a projective plane will be described systematically; the construction leading to the projective space is quite similar and only the major differences will be presented. As shown later, there exist very good means to give an intuitive picture of a projective plane whereas it is much more difficult to visualise a projective space; consequently, projective planes will always be used as illustrative examples even if the real environments used in graphics systems tend to be a projective space rather than a plane.

If $\mathbb{E}^2$ is the Euclidean plane, let us denote by $\Lambda(\mathbb{E}^2)$ the set of all lines of $\mathbb{E}^2$. On this set, the relation of *parallelism* is an equivalence relation (provided that each line is considered to be parallel to itself). In other words, the relation ρ could be defined by:

$$\forall n,m \in \Lambda(\mathbb{E}^2): n\rho m \leftrightarrow n \text{ and } m \text{ are parallel} \tag{2.8}$$

The fact that this relation is an equivalence relation can be seen easily. Consequently, one may speak of the quotient space of $\Lambda(\mathbb{E}^2)$, which (instead of $\Lambda(\mathbb{E}^2)_\rho$) will be simply denoted by $\mathit{\Pi}$.

Intuitively speaking the elements of $\mathit{\Pi}$ are mathematically precise abstractions of what is common in two lines of $\mathbb{E}^2$ vis-a-vis parallelism. Indeed, two parallel lines will generate the very same element of $\mathit{\Pi}$ (using formula (2.6)) and elements generated by two non-parallel lines will be different. This also means that if a new set is defined by

$$\mathbb{PE}^2 = \mathbb{E}^2 \cup \mathit{\Pi} \tag{2.9}$$

the resulting set will contain on the one hand the "traditional" Euclidean points plus some abstract elements describing the "common part" of two parallel lines of $\mathbb{E}^2$.

To define some kind of geometry on $\mathbb{PE}^2$, the notions of points, lines, intersections of lines etc. have to be extended onto this larger set; of course only an extension which would preserve the "traditional" Euclidean notions is of real interest. By defining these extensions and by finding some elementary properties of them, a new mathematical structure, a new *geometry* will be created. Mathematically, this means that the set of elementary properties might also be considered as a new set of axioms (much the same way as the axioms described in the Elements of Euclid form the basis of Euclidean geometry or the modified set of these axioms defined by Bólyai and Lobatchewsky would form the axiomatic basis for hyperbolic geometry). This new geometry is called *projective geometry*.

In Euclidean geometry, the notion of point is just another name for the (set-theoretical) elements of the set $\mathbb{E}^2$ (or $\mathbb{E}^3$). The same approach can be used in the case of projective geometry, that is:

Definition 2.1. The elements of $\mathbb{PE}^2$ are called *(projective) points*. In case it is necessary to make a difference, the elements of $\mathit{\Pi}$ are also called *ideal points* whereas the elements of $\mathbb{E}^2$ are also called *affine points[†]*.

Lines in Euclidean geometry are special subsets of $\mathbb{E}^2$; the properties of these subsets are described in the axioms of the Euclidean axiomatic system (the notation $\Lambda(\mathbb{E}^2)$ has been used to denote the set of all lines). The aim is to maintain Euclidean lines in the new environment as well. Here is the precise definition:

[†]The term *directions* is also in use for affine points.

Definition 2.2. *Lines* of IPE^2 are special subsets of IPE^2. The set of all lines is denoted by $\Lambda(IPE^2)$ and its elements can be described as follows:

$$\Lambda(IPE^2) = \{x \cup \{x_\rho\} : x \in \Lambda(IE^2) \} \cup \{II\}$$

where ρ stands for the equivalence relation "parallelism".

In plain English: each line of IE^2 is extended by its direction, that is the ideal point generated by the line using (2.6); additionally, the set of all ideal points is also considered as a line. Just as in the case of points, if there is a necessity to make a difference, II will also be called the *ideal line* (there is only one such line!) whereas all other lines are the *affine lines*. An affine line is *not* equal to a Euclidean line; it is a Euclidean line *plus* one point (also called the ideal point of the line).

The intersection of a line and a point is just another terminology for the set–theoretical inclusion; it is therefore automatically valid for the new environment as well.

What are the basic properties of these lines and points (also called, to make the distinction, projective lines and points)? There are some statements forming a set of basic theorems on these notions and which are as follows.

Theorem 2.1. For each two elements of IPE^2 the following holds:

$$\forall P,Q \in IPE^2, (P \neq Q): \exists! \ l \in \Lambda(IPE^2) \text{ for which}$$

$$P \wedge l \neq \emptyset \text{ and } Q \wedge l \neq \emptyset$$

That is for every pair of projective points there exists one and only one projective line which contains the given two points.

The proof of this theorem follows a fairly standard mathematical line of thought: different cases should be examined apart.

(1)   If both P and Q are affine points, the corresponding axiom of the Euclidean geometry says that there is one and only one Euclidean line which intersects both P and Q; the corresponding affine line will do for the projective case. It is trivial to see that no other projective line will satisfy the requirements.

(2)   If both P and Q are ideal points, the ideal line will contain both of them; furthermore, no affine line may intersect two distinct ideal points.

(3)   Finally if P is affine and Q ideal, there is a whole set of affine lines intersecting P. However, out of these lines only one may have as an ideal point Q: the ideal point is just an element of II determined by the relation (2.6). ■

An analogous statement for lines is as follows.

Theorem 2.2. For every pair of lines on $I\!PE^2$ the following holds:

$$\forall l, n \in \Lambda(I\!PE^2)\ (l \neq n)\colon \exists!\ P \in I\!PE^2 \text{ for which}$$

$$P \wedge l \neq \varnothing \text{ and } P \wedge n \neq \varnothing$$

That is each two distinct lines have an intersection point (there are no parallel lines!). This intersection point is denoted by $l \wedge n$. The proof of this theorem is similar to the previous one:

(1) If both n and m are affine, there are again two cases:

   (a) the two lines are parallel in the Euclidean sense; in this case they share the same ideal point (according to (2.6));

   (b) the two lines have an Euclidean intersection point which will also serve as an intersection point in the projective sense as well.

(2) If n is affine and m is the ideal line, the ideal point of n (which exists according to definition 2.2) is the intersection point. ∎

Finally, two theorems are needed which are rather technical in nature but are necessary for the full description of the whole theory:

Theorem 2.3. Each projective line contains at least three points.

Theorem 2.4. There exist three points on $I\!PE^2$ which are not on the same line (they are not *collinear*).

Theorems 2.1 to 2.4, together with the corresponding definitions 2.1 and 2.2 form the axiomatic foundation of (planar) projective geometry. By using these notions and theorems the ambiguities of the description related to figure 2.1 may now be removed. The planes Π and Ψ should now be considered projective planes instead of Euclidean ones. P, which is an affine point of Ψ, will be mapped onto an ideal point of Π; the lines n and m which intersect at P on Ψ will be mapped onto Π (that is onto n' and m') by still keeping the line intersection; the only problem is that this intersection point happens to be an ideal point. In the case of projective geometry, this makes no real difference, however.

A *projective space* can be constructed very similarly. Instead of $\Lambda(I\!\!E^2)$, $\Lambda(I\!\!E^3)$ should be considered for the equivalence relation; the resulting quotient set will contain the set of *ideal points* again. The set of all projective points will be

$$I\!PE^3 = I\!\!E^3 \cup I\!I \tag{2.10}$$

just as in the case of the plane.

The definitions corresponding to 2.1 and 2.2 are very similar but there are however some differences. In the case of spatial geometry, there are two special kinds of subspaces: lines and planes. Likewise, projective lines and projective planes should both be defined to ensure a correct extension.

The most important remark which helps to make these extensions possible is

14

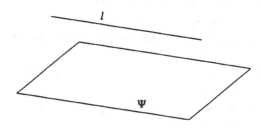

Figure 2.2.

as follows. If $\Pi(I\!\!E^3)$ denotes the set of all planes of $I\!\!E^3$ and $\Psi{\in}\Pi(I\!\!E^3)$, this plane generates a special subset of $I\!\!I$ (see also figure 2.2). Indeed, the ideal points which are parallel to Ψ (like the one generated by l on figure 2.2) will have a special characteristic: they will describe all the planes and lines which are parallel to Ψ. Denoting this subset of $I\!\!I$ by Ψ_ρ, the following definition may be accepted:

Definition 2.3. *Lines* of $I\!\!PE^3$ are special subsets of $I\!\!PE^3$. The set of all lines is denoted by $\Lambda(I\!\!PE^3)$ and its elements may be described as follows:

$$\Lambda(I\!\!PE^3) = \{x{\cup}\{x_\rho\} : x{\in}\Lambda(I\!\!E^3)\} \cup \{\Psi_\rho : \Psi{\in}\Pi(I\!\!E^3)\}$$

In other words, each spatial plane induces a subset of $I\!\!I$ which is called an *ideal line*; there is one such ideal line for each set of parallel planes. It is intuitively clear (and mathematically easily provable) that the equivalence classes induced on $\Pi(I\!\!E^3)$ by the relation parallelism are isomorphic to these ideal lines.

Definition 2.4. *Planes* of $I\!\!PE^3$ are special subsets of $I\!\!PE^3$. The set of all planes is denoted by $\Pi(I\!\!PE^3)$ and its elements may be described as follows:

$$\Pi(I\!\!PE^3) = \{\Psi{\cup}\{\Psi_\rho\} : \Psi{\in}\Pi(I\!\!E^3)\} \cup \{I\!\!I\}$$

Here again, $I\!\!I$ is called the *ideal plane* whereas all other planes are denoted by *affine planes*. Definitions 2.1, 2.3 and 2.4 form the necessary extensions for projective space. There is a set of theorems which form the necessary axiomatic basis for the theory. The theorems themselves are very much the same as above in flavour and they are just listed here without proof (the reader may easily reproduce the mathematical deductions). These theorems are as follows (some very technical ones like the analogies of theorem 2.4 are omitted).

Theorem 2.5. If a point is the element of a line and the line is a subset of a plane then the point is an element of the plane (that is the two different kind of subsets in $I\!\!PE^3$ form a "hierarchy").

Theorem 2.6. Every pair of points generate one and only one line which contains them both (denoted by $P \vee Q$).

Theorem 2.7. Every pair of planes have one and only one common intersection line (denoted by $\Pi \wedge \Psi$). In other words, there are no parallel planes.

Theorem 2.8. A line and a plane has either an intersection point or the line belongs to the plane.

Theorem 2.9. For every pair of lines there is at most one plane which contains them both (denoted by $l \vee n$).

Theorem 2.10. Each three non collinear points generate one and only one plane which contains them all (denoted by $P \vee Q \vee S$).

Theorem 2.11. Each three planes intersect in either a line or one point (denoted by $\Pi \wedge \Psi \wedge \Phi$).

Theorem 2.12. If two lines are coplanar (that is there exists a plane which contains them both), they have one and only one intersection point (denoted by $l \wedge m$).

One has to be very careful in the case of the last statement: in projective space there may be lines which do not have an intersection point (these kinds of lines are not considered to be parallel in classical Euclidean geometry either). It is true, however, that there are no parallel lines any more. Another point of interest: each plane in $I\!P E^3$ can be considered as a projective plane by itself (just as each plane in $I\!\!E^3$ behaves "locally" as a Euclidean plane). This statement sounds trivial but in a precise formulation of projective geometry it has to be proven that the axioms of the projective plane are valid locally as well.

As said before, theorems 2.1 up to 2.4 together with the definitions 2.1 and 2.2 (or their three–dimensional counterparts), may be considered as an axiomatic system: this is the axiomatic foundation of projective geometry. In theory, based on these axioms, all the (sometimes obscure) steps of the construction may be relegated to the background: one could just speak of points and lines in $I\!P E^2$ where no parallel lines exist any more. This geometry has a very different nature compared to the "well–known" Euclidean one. A projective plane or space is locally very much like its Euclidean counterpart but has different global characteristics (the exact relationships between Euclidean and projective geometry will become clear in a later chapter).

The main difference comes from the fact that a projective line behaves much like an Euclidean (planar) circle; in fact, it is isomorphic with it. The ideal point of the line is the element which somehow "glues" the two ends. This fact has far reaching consequences: the very notion of line segment has no meaning any more (by giving the points A and B on the line, one can reach B from A in two ways). Consequently, the concept of the interior of a polygon disappears from the theory as do convex polygons.[†] There is no way of defining the notions of "clockwise" and

---

[†]More generally, Jordan's theorem, which states the existence of the interior and the exterior of a planar area generated by a "well-behaved" curve on the Euclidean plane is not true any more.

"anticlockwise" on a projective plane; mathematically speaking, the projective plane is not orientable.

Of course, these differences are at the source of a number of problems when projective geometry has to be used for the purposes of computer graphics. Computer graphics applications rely heavily on, for example, the interior of a polygon, which also forms an integral part of all ISO standards on graphics (see all ISO documents in the references). One possibility would be to avoid the use of this theory; however, as stated in the introduction, this is barely possible. Another approach would be to develop an alternative mathematical theory which would try to avoid the appearance of these problems; an attempt has been made recently by J. Stolfi ([Stol89]) based on some earlier mathematical works made by H. Grassmann about a hundred years ago and followed by a number of other mathematicians (Stolfi refers to the works [Berm61] and [Hest84] for earlier references). In his thesis, Stolfi describes the theory of *oriented projective spaces*, which contains many similarities to classical projective geometry but where the notion of the orientation of a line, plane and space still has a meaning. However, the mathematical theory and formulae involved tend to be rather complicated and quite abstract; to use it in practice would probably require a reformulation of a number of classical approaches which have been in use in computer graphics in the past 10 to 15 years. Whether this is worthwhile or not is still to be proven; the approach is of interest, however.

The approach described in this thesis is much more pragmatic. Projective geometry should be used, because it is a precise description of practical problems arising in computer graphics and it also helps to create more efficient algorithms and methods. However, the extreme axiomatic nature of projective geometry, which would ignore the origins of, for example, ideal points will not be followed everywhere; in most of the cases the construction described here will be present in the background. By carefully exploiting the relationships between projective geometry and the Euclidean one, some of the problems may be described in terms of a Euclidean environment even if the price to be paid might be sometimes to use four dimensional geometry instead of the well known two and/or three dimensional ones. In some other cases the full power of projective geometry has to be used (eg for the handling of conics). By alternating between projective and Euclidean geometries, most of the problems can be avoided in a down–to–earth but still powerful way.

In contrast to the traditional and "purist" projective geometry textbooks, the numerical aspects of projective geometry are of primary interest to computer graphics scientists. Some kind of coordinate system is essential to be able to describe geometric entities with numbers and hence make them manageable by computers. This is will be covered in the next section.

2.4. Projective Coordinate Systems (Homogeneous Coordinates)

Coordinate systems as used in Euclidean geometry were only introduced in the 18th century. Their use has become so natural that one tends to underestimate the importance and the mathematical difficulties involved when using them. The use of the

Cartesian system creates a "bridge" between two very different mathematical theories, namely Euclidean geometry and the theory of real numbers. A more exact mathematical formulation of what the Cartesian coordinate system really means is presented here, to show what is the necessary approach to achieve something analogous for projective plane/space.

Theorem 2.13. If O, E_1 and E_2 are three non-collinear points of the Euclidean plane $I\!E^2$, then there exists a one–to–one correspondence between $I\!E^2$ and $I\!R^2$ so that the point O will correspond to the vector $(0,0)^T$, the point E_1 to $(1,0)^T$ and, finally, E_2 to $(0,1)^T$. If, furthermore, it is required that the distance of the points P and Q should be expressed by the formula:

$$dist(P,Q) = \sqrt{(p_1 - q_1)^2 + (p_2 - q_2)^2}$$

(where P is mapped onto $(p_1,p_2)^T$ and Q onto $(q_1,q_2)^T$), then there exists only one such correspondence which fulfils these requirements.

It is not possible to have a one–to–one correspondence between the projective plane and $I\!R^2$. A mapping is however provided by the use of *homogeneous coordinates*.

Two non–zero vectors $a,b \in I\!R^n$ are considered to be equal in the homogeneous sense if there exists a non–zero $\lambda \in I\!R$ so that the equality $a = \lambda b$ holds. This relation is an equivalence relation on $I\!R^n - \{0\}$ (the origin). The corresponding quotient set (denoted by $I\!P\!R^n$ in the following discussion) is called the set of homogeneous vectors. The same vector notations will be used to denote its elements, but their homogeneous nature must always be kept in mind. In case of doubt, the notation $[(a_1,a_2,\ldots,a_n)]^T$ will also be used to denote the homogeneous vector generated by $(a_1,a_2,\ldots,a_n)^T \in I\!R^n$.

The theorem which is analogous to 2.13 is as follows.

Theorem 2.14. If O, A_1, A_2 and E are elements of $I\!P\!E^2$, so that no three of them would be collinear (they are of a *general position*), then there exists a unique one–to–one correspondence between the points of $I\!P\!E^2$ and the elements of $I\!P\!R^3$ so that the following relations

$$O \leftrightarrow [(0,0,1)]^T$$
$$A_1 \leftrightarrow [(1,0,0)]^T$$
$$A_2 \leftrightarrow [(0,1,0)]^T \tag{2.11}$$
$$E \leftrightarrow [(1,1,1)]^T$$

hold.

For a projective space, one more point (A_3) is necessary; the requirement is not only that there should be no three collinear points but also that there should not be four coplanar points. The mapping is then performed between $I\!P\!E^3$ and $I\!P\!R^4$ and (2.11) becomes:

$$
\begin{aligned}
O &\leftrightarrow [(0,0,0,1)]^T \\
A_1 &\leftrightarrow [(1,0,0,0)]^T \\
A_2 &\leftrightarrow [(0,1,0,0)]^T \\
A_3 &\leftrightarrow [(0,0,1,0)]^T \\
E &\leftrightarrow [(1,1,1,1)]^T
\end{aligned}
\tag{2.12}
$$

Traditionally, the coordinate components of the homogeneous vectors are denoted either by subscripted Latin characters or by using the letter w for the last coordinate eg of the form $[(x,y,w)]$. The exact proofs of the theorems 2.13 and 2.14 would go far beyond the scope of the present thesis; the interested reader should consult, for example, [Keré66].

The exact relationships between the Cartesian coordinates and the homogeneous ones play an essential role in the following sections. Indeed, the geometric environment which is the usual starting point for any graphics system is a Euclidean one together with some coordinate system defined on $I\!E^2$ or $I\!E^3$; a coordinate system which would be effective in some sense for the object which is to be described. This Euclidean environment has to be treated, however, as a projective one by the graphics system; there is therefore an extension to be made (described in the previous section) which would embed this plane or space into a projective plane or space respectively. This process is performed by adding ideal points to $I\!E^2$ or $I\!E^3$ to result in $I\!P\!E^2$ or $I\!P\!E^3$ respectively. The question is, which homogeneous coordinate system to use so that the relationship between the Cartesian and the homogeneous coordinates of a Euclidean (that is affine) point would be as simple as possible? This is done as follows.

Theorem 2.15. Suppose that a Cartesian coordinate system has been chosen on $I\!E^2$ (for the sake of simplicity the planar situation will be examined in detail first). Let the points O, A_1, A_2 and E be defined as follows.

- Let O be the origin of the Cartesian system;
- Let A_1 be the ideal point of the x axis;
- Let A_2 be the ideal point of the y axis;
- Let E be the affine point with coordinates $(1,1)^T$

then, if the Cartesian coordinates of a point P are denoted by $(p_1,p_2)^T$ the following relation holds:

If $P \in I\!E^2$, the homogeneous coordinates of P are given by $[(p_1,p_2,1)]^T$.

Furthermore, if Q is an ideal point of $I\!PE^2$, it can be described by the line $Q \vee O$. With the help of such a line the following relationship also holds:

If $R \in Q \vee O$, the homogeneous coordinates of Q are given by $[(r_1, r_2, 0)]^T$.

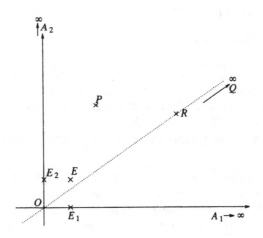

Figure 2.3.

It is fairly straightforward to see that the relations listed in theorem 2.15 do define a one–to–one mapping of $I\!PE^2$ and the set of homogeneous vectors. Taking into account the uniqueness statement formulated in theorem 2.14, the validity of theorem 2.15 follows easily. ∎

Theorem 2.15 also means that *in this coordinate system* ideal points can be uniquely characterised by having the last coordinate value being zero. This also leads to the following formula (well known in computer graphics, see also figure 2.3):

If $P \in I\!PE^2$, P is affine and a homogeneous coordinate system has been chosen for $I\!PE^2$ as described above, the formula

$$[(p_1, p_2, p_3)]^T \rightarrow (p_1/p_3, p_2/p_3)^T \qquad (2.13)$$

will give the Cartesian coordinate values of P (the fact that P is affine is equivalent to the fact that $p_3 \neq 0$). In the computer graphics literature this step is usually called the "projective division"[†].

It has to be stressed that such a unique characterisation of an ideal point by its

---

[†]The terms "perspective division" or "w-divide" are also in use.

coordinate values is possible only in this coordinate system. It is perfectly possible to choose another homogeneous coordinate system where this characterisation is not valid any more[†]. Just as in the case of Cartesian coordinate systems, it is a fairly widespread approach, when trying to prove some theorem in projective geometry, to choose a coordinate system which fits the original problem itself (this approach is particularly fruitful when dealing with conics).

Homogeneous coordinates also provide a way of characterising lines (on IPE^2) and planes (on IPE^3). In the case of Cartesian coordinates, a line can be expressed by the equation:

$$ax_1 + bx_2 + c = 0 \qquad (2.14)$$

with appropriate constants a, b and c. By using the identification procedure for Cartesian versus homogeneous coordinates, this equation could be rewritten for homogeneous coordinates as follows:

$$ax_1 + bx_2 + cx_3 = 0 \qquad (2.15)$$

It is also clear that if the values a, b and c are multiplied by any non–zero real number, equation (2.15) would still remain valid. This means that $[(a,b,c)]^T \in IPR^3$ gives an adequate description of a line. In other words, by using homogeneous coordinates, not only points but also lines can be assigned a homogeneous vector, which could just be called the homogeneous coordinates of a line (and the points of the line can be described by (2.15)). In case of IPE^3 planes can be described similarly (but not lines).

It can be proven that the above characterisation of lines/planes does not depend on the special choice of the homogeneous coordinate system used to derive (2.15). The fact that lines/planes can be described by homogeneous coordinates just like points leads to an elegant symmetry of all the formulae involving intersection points, generated lines etc.; they will be presented later. This similarity, which is referred to as the *duality principle* of projective geometry has, in fact, a much deeper background. By looking at the axioms of IPE^2 (or respectively of IPE^3), there is a similarity of the behaviour of points and lines: if the words/terms "point" are exchanged with "line", and "intersection points of lines" with "lines generated by points", valid statements will result. Consequently, this fact is also true not only for the axioms but for all statements derived from them. By making use of this duality principle a number of formulae can be derived easily and one might also get a clue for finding additional and useful formulae (see for example [Arok89]).

The use of homogeneous coordinates has been an accepted practice in computer graphics for a very long time; their description can be found in all "classical"

[†] In fact, the whole approach might also be turned upside down. Indeed, if an arbitrary homogeneous coordinate system is given on IPE^2 or IPE^3, one could *define* the ideal points *of this coordinate system* to be the points with the last coordinate value being zero and all other ones being affine.

textbooks ([Newm79], [Fole84], [Salm87], [Fole90] and others) and more systematic descriptions can also be found in [Reis81] or [Bez83]. In most of these cases however, homogeneous coordinates are presented as being some kind of neat (one could even say "tricky") way of describing points so as to have a unified description of the effect of different transformations. While the usability of homogeneous coordinates even in 2D graphics is undeniable, it is important to realise that their use can be traced back to much more fundamental mathematical properties of projective geometry, which, in turn, plays a basic role in 3D graphics.

2.5. Isomorphic Models of Projective Planes and Projective Spaces

The use of homogeneous coordinates provides a means of "visualising" a projective plane (and to a smaller extent a projective space). The idea is to give some kind of an intuitively manageable surface in Euclidean geometry which would, in some sense, be a good model for a projective environment.

A homogeneous coordinate (that is an equivalence class) can be viewed as a line in a higher dimensional Euclidean space. That is, an element of $I\!PR^3$ can be identified with a line in $I\!R^3$ crossing the origin[t]. This fact is clear from the definition of a homogeneous coordinate.

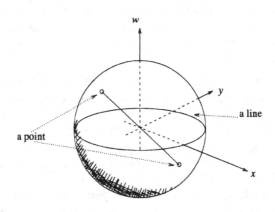

Figure 2.4.

2.5.1. The Riemann Sphere

The Riemann sphere (figure 2.4; also called the spherical model of a projective plane) is the unit sphere around the origin of $I\!R^3$ where all diametrically opposite points are identified. Each point in $I\!PE^2$ is represented by a poins of the Riemann sphere, and lines of $I\!PE^2$ are represented by the great circles on the sphere (again,

---

[t] More precisely, the origin should be removed from the line to get the exact identification.

with opposite points identified). The ideal line is represented by the great circle defined by $w=0$. With this mapping of $I\!PE^2$ onto the Riemann sphere, the latter becomes isomorphic to $I\!PE^2$. The model also shows the remarkable identity of an affine line and the ideal line.

The Riemann sphere is, although intuitively very helpful, not really of use in the forthcoming. The reason is that the identification of a sphere point with the corresponding affine point of $I\!PE^2$, and also with its Cartesian coordinates, leads to disagreeable formulae. It is, however, a very helpful tool to get new ideas; indeed, as opposed to the so called "straight model" of the projective plane (presented in the next section) the full plane, including the ideal line, is modelled by it.

The Riemann sphere can be generalised for projective space as well; one should take a unit sphere in $I\!R^4$ around the origin However, because of the difficulties of visualising a four dimensional space, this version of the Riemann sphere is not really useful.

2.5.2. Embedding into $I\!R^3/I\!R^4$ (the Straight Model)

The second, and more widespread model of a projective plane (the so–called straight model) is shown in figure 2.5. By using the identification of Cartesian and homogeneous coordinates, all *affine* points (that is the points of $I\!E^2$) may be represented by points of Π (that is the plane $w=1$). This is clearly nothing else than a pictorial representation of the fact that homogeneous coordinates describe lines in a higher dimensional space. From a purely mathematical point of view, the drawback of this model is the fact that only the affine points can be represented so clearly. On the other hand, as far as computer graphics is concerned, the real issue is always to see how affine points are transformed; ideal points are just disagreeable but necessary additions to them. In this sense, the fact that the straight model shows only the affine points so clearly may well be an advantage rather than a disadvantage.

On the straight model ideal points are represented by homogeneous coordinates with the last coordinate value being zero. This means that ideal points are represented by lines running in the plane $x-y$, crossing the origin and having therefore no intersection points with Π (see figure 2.5 again).

The fact of having chosen Π to represent the projective points was, although a direct representation of the Cartesian–homogeneous identification, intentional: as said above, the fact that ideal points are so well separated from the affine ones might be helpful. Clearly, any plane could have been chosen equally well, like the plane Ψ in figure 2.6. In this case the ideal points do become Euclidean points on Ψ but, on the other hand, some of the affine points cannot be represented properly.

Like the Riemann sphere, this model works analogously for $I\!PE^3$: one should take the $w=1$ hyperspace in $I\!R^4$ to model $I\!PE^3$. Of course, the same problem arises: it is not possible to visualise properly the four dimensional space. This is the reason why figures 2.5 or 2.6 will be used in all cases when a pictorial representation will be necessary even if the real problems to be solved will be in $I\!PE^3$ rather than in $I\!PE^2$.

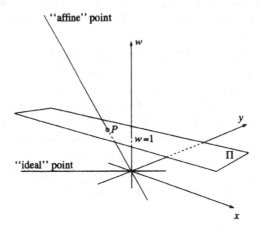

Figure 2.5.

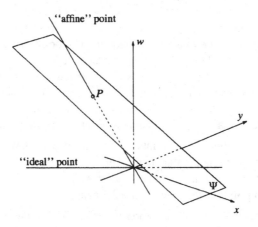

Figure 2.6.

There is one common point which has to be stressed in all forms of the straight model. In all these cases the projective environment has been embedded into a Euclidean environment again (with some restrictions). The significant difference is, however, that a Euclidean space of higher dimension was necessary. This will have an importance in what follows: by having first embedded the original Euclidean environment into a projective one and, in a second step, having re-embedded it into a Euclidean space of a higher dimension, many of the problems can be restated in a "classical" Euclidean way. This fact has never been really exploited in classical projective geometry; indeed, the problems arising for mathematicians are of a different nature. However, for computer graphics, this approach (exploited first in [Herm87]) has led to significant simplifications of a number of problems. Examples will be seen in later chapters.

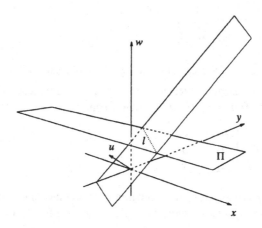

Figure 2.7.

The straight model also provides a way to represent all affine lines or planes (see figure 2.7). Let $u \in IPR^3$ (or, for IPE^3, $u \in IPR^4$) be the homogeneous representation (that is the coordinates) of the line l. As an Euclidean vector, u will also determine a plane in IR^3 (resp. a hyperspace in IR^4), namely the one containing the origin and whose normal vector is u. This plane (hyperspace) will intersect Π in a line (resp. a plane) if l is affine; the formula describing a line in homogeneous coordinates (that is (2.15)) simply proves that this intersection line/plane will just be l itself. In other words, each affine line can be viewed as a plane (or a hyperspace) crossing the origin in IR^3 (resp. IR^4) if the straight model is used.

2.6. Some Basic Calculation Formulae

In what follows, a number of formulae (for the intersection of lines etc.) will be presented; these formulae will be used later. This is by no means an exhaustive list of all possible calculation methods offered by homogeneous coordinates; the reader should refer to the projective geometry textbooks and primarily to [Penn86] for further examples.

2.6.1. Formulae for the Projective Plane

If $u \in IPR^3$ and $v \in IPR^3$ represent two lines, the homogeneous coordinates of $u \wedge v$ are given by the following (formal) determinant:

$$u \wedge v = \det \begin{bmatrix} u_1 & u_2 & u_3 \\ v_1 & v_2 & v_3 \\ e_1 & e_2 & e_3 \end{bmatrix} \qquad (2.16)$$

where e_1, e_2 and e_3 denote the vectors $(1,0,0)^T$, $(0,1,0)^T$ and $(0,0,1)^T$ respectively.

If $p \in IPR^3$ and $q \in IPR^3$ represent two points, the homogeneous coordinates of $p \vee q$ are given by the following (formal) determinant:

$$p \vee q = \det \begin{bmatrix} p_1 & p_2 & p_3 \\ q_1 & q_2 & q_3 \\ e_1 & e_2 & e_3 \end{bmatrix} \qquad (2.17)$$

To describe $p \vee q$, a parametric equation is sometimes more suitable than the description with homogeneous coordinates. The following formula is also true:

$$p \vee q = \{ \lambda p + \mu q \} \qquad (2.18)$$

$$\lambda \neq 0 \text{ or } \mu \neq 0$$

The previous formulae may be used for slightly more complex calculations; for example to find the intersection point of two lines, knowing two points on each of them (a repetitive application of (2.17) and then (2.16) will do).

It is also important to know that in the case of IPE^2 the coordinates of some special points are known "by default". As examples, at least two distinct ideal points are known by their coordinates (eg $[(1,0,0)]^T$ and $[(0,1,0)]^T$), the homogeneous coordinate of the ideal line is also known (it could be calculated by using (2.17) and the two distinct ideal points but it is also clear that $[(0,0,1)]^T$ should be the result).

It can be of interest to see what the ideal point of a given line is, provided that two of its affine points, say $p \in IPR^3$ and $q \in IPR^3$, are given. This can easily be calculated if the "straight model" is made use of: indeed, the direction of a line parallel to Π and the (Euclidean!) line $p' \vee q'$ is to be found. However, if p and q are considered to be Euclidean, the vector from q' to p' can be calculated by:

$$\begin{bmatrix} p_1/p_3 \\ p_2/p_3 \\ 1 \end{bmatrix} - \begin{bmatrix} q_1/q_3 \\ q_2/q_3 \\ 1 \end{bmatrix} \qquad (2.19)$$

Clearly, the homogeneous coordinates generated by (2.19) will give the ideal point of $p \vee q$ (see also figure 2.8).

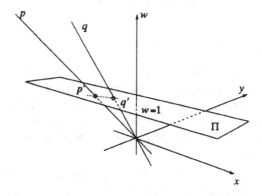

Figure 2.8.

2.6.2. Formulae for the Projective Space

If $u \in I\!PR^4$, $v \in I\!PR^4$ and $w \in I\!PR^4$ represent three planes, the homogeneous coordinates of $u \wedge v \wedge w$ are given by the following (formal) determinant:

$$u \wedge v \wedge w = \det \begin{bmatrix} u_1 & u_2 & u_3 & u_4 \\ v_1 & v_2 & v_3 & v_4 \\ w_1 & w_2 & w_3 & w_4 \\ e_1 & e_2 & e_3 & e_4 \end{bmatrix} \tag{2.20}$$

where e_1, e_2 e_3 and e_4 denote the vectors $(1,0,0,0)^T$, $(0,1,0,0^T)$, $(0,0,1,0)^T$ and $(0,0,0,1)^T$ respectively. If all subdeterminants in (2.20) are zero, the planes meet in a line; this is a singular case.

If $p \in I\!PR^4$, $q \in I\!PR^4$ and $r \in I\!PR^4$ represent three points, the homogeneous coordinates of $p \vee q \vee r$ are given by the following (formal) determinant:

$$p \vee q \vee q = \det \begin{bmatrix} p_1 & p_2 & p_3 & p_4 \\ q_1 & q_2 & q_3 & q_4 \\ r_1 & r_2 & r_3 & r_4 \\ e_1 & e_2 & e_3 & e_4 \end{bmatrix} \tag{2.21}$$

It is also worth mentioning that formula (2.18) remains valid in $I\!PE^3$ as well, although the fundamental symmetry among formulae is now between points and planes instead of points and lines.

In $I\!PE^3$ at least three distinct and non–collinear ideal points are known by their coordinates (eg $[(1,0,0,0)]^T$, $[(0,1,0,0)]^T$, $[(0,0,1,0)]^T$). The homogeneous vector $[(0,0,0,1)]^T$ gives the coordinates of the ideal plane.

If three non-collinear points of a plane Π are known, the homogeneous representation of two ideal points of Π can be calculated analogously to (2.19). By using (2.18), the parametric representation of the ideal line on Π can also be described.

2.7. Collinearities

In his already cited work, the *Erlanger Programm*, F. Klein has proposed the classification of different branches of geometry based on a class of transformations which would leave some part of the geometric structure invariant. Although the usual geometry textbooks do not follow this rigorous classification, to describe the properties of a geometry the description of an appropriate class of transformation might be very important.

In the case of projective geometry the geometrical structure is fully determined by lines, the intersection of lines and of the lines generated by points. It is therefore natural to accept the following definition to describe a basic set of transformations:

Definition 2.5. A transformation $T: I\!PE^2 \to I\!PE^2$ (or $T: I\!PE^3 \to I\!PE^3$) is said to be a collinearity if for each three collinear points $P, Q, R \in I\!PE^2$ (or $P, Q, R \in I\!PE^3$ respectively) the points $T(P), T(Q), T(R)$ are also collinear.

Collinearities (also called *projective transformations* or *projective mappings*) play a significant role in projective geometry. These transformations map lines onto lines (according to definition 2.5), they map the intersection point of two lines onto the intersection point again and, finally, they also map the lines generated by points onto the line generated by the images of the points. In case of $I\!PE^3$, planes are also mapped onto planes as well as the intersection of planes onto the intersection of planes.

The theoretical importance of collinearities can be well understood from the fact that they keep invariant all properties used and described by the axioms of projective geometry. Indeed, all axioms are specified with the help of points, lines and intersections, properties which are invariant to projective mappings. The practical importance of these mappings is also high: "familiar" transformations like rotations, translations or scalings as well as central projections are all collinearities. Care should be taken, however, with the last example. Central projections are collinearities in the *projective sense* whereas they are not necessarily ones in the Euclidean sense; this was exactly the problem which had led to the use of this theory. Those readers who are familiar with the three dimensional graphics standards like GKS–3D, PHIGS ([ISO88], [ISO89], [ISO89a]), or related packages like PEX ([ISO88b], [Clif88]) can also realise that the notion of projective transformations encapsulates all possible transformations described in these documents, like modelling transformations and viewing. These examples will be detailed in what follows.

For practical and also theoretical purposes a very important sub–class of collinearities is the class of *affine transformations*. Its definition is as follows.

Definition 2.6. A collinearity is said to be an *affine* transformation if the images of all affine points remain affine. For non–singular transformations this also means that the images of all ideal points remain ideal.

Clearly, affine transformations are "closer" to Euclidean geometry than projective transformations in general; they leave the "dual" structure of a projective plane/space (ie the division among ideal and affine points) essentially intact. Among the examples cited above, rotations, translations and scalings are clearly affine while central projections are not.

A line (for $I\!PE^2$) or a plane (for $I\!PE^3$) is called the *vanishing line/plane* of the transformation if it is transformed onto the ideal line/plane respectively. A transformation is affine if and only if its vanishing line/plane is the ideal one.

One very important issue in projective geometry is to find the *projective invariant* features of geometric primitives or constructions. Projective invariance means that the given construction and/or the geometric features related to a given primitive would remain unchanged if a projective transformation were applied (more specific examples will be given later). Likewise, one could speak about *affine invariance*, related to features invariant for affine transformations.

Projective and affine invariance are not only of theoretical interest. In the case of computer graphics algorithms, one of the clues for a simplification or an improvement might be to find the projective/affine invariant part of them. As an example (and there will be much more later) one might think of the fact that a number of graphics primitives may be described in a very compact form with the help of some points only (eg conics), but for the final rendering some kind of a linear approximation of the primitive is necessary. It is of great importance to find projective/affine invariant representations of these primitives; such representations allow the postponement of the linear approximation along the graphics output pipeline, resulting therefore in faster rendering and better approximations.

The existence and uniqueness of collinearities is provided by the following theorem.

Theorem 2.16. If Π and Ψ are two projective planes (which may be identical), $P_i \in \Pi$ and $P_i' \in \Psi$ ($i=1,...,4$) are points in Π and Ψ respectively such that no three of them are collinear, then there exists one and only one collinearity $T: \Pi \rightarrow \Psi$ for which $T(P_i) = P_i'$. For projective spaces five points are needed instead of four with the additional requirement that no four points may be coplanar.

Here again, the proof of this theorem would go beyond the scope of this thesis; the interested reader should consult for example [Keré66], [Coxe49] or [Penn86].

This theorem seems to have a theoretical value only, as far as computer graphics are concerned, but this is not absolutely true. It happens quite often that several, at first glance very different, methods are created to calculate an effective projective transformation. Theorem 2.16 provides a way to check whether the different approaches do generate the same mapping or not: only the images of four/five points have to be checked and the theorem ensures the uniqueness of the projective mapping provided that the images of these four/five points coincide.

2.7.1. Representation of Collinearities

As with projective points it is obviously of paramount importance to find some kind of numerical representation of collinearities. The corresponding theorem (which is one of the most important theorems in projective geometry) is as follows.

> **Theorem 2.17.** If Π and Ψ are two projective planes (which may be identical) with a homogeneous coordinate system chosen on them and $T: \Pi \rightarrow \Psi$ is a collinearity, then there exists a 3×3 matrix $\bar{T}$ which describes the transformation as follows. If $x \in \mathit{IPR}^3$ represents the point $X \in \Pi$ then
>
> $$[\bar{T}x] \in \mathit{IPR}^3 \qquad\qquad (2.22)$$
>
> gives the homogeneous coordinates of $T(X)$. Furthermore, $\bar{T}$ is uniquely defined in a homogeneous sense; that is, if $\bar{T}$ and $\bar{T'}$ both fulfil (2.22), then there exists a non-zero real number λ for which $\bar{T} = \lambda \bar{T'}$.

In other words, the transformation can be described by a matrix-vector multiplication. For IPE^3, the same theorem applies, with the obvious difference that the matrices involved are 4×4 rather than 3×3. In both cases, the singularity of the transformation is equivalent to the singularity of the corresponding matrix.

The opposite statement is also true, namely that formula (2.22) defines a collinearity for all 3×3 (respectively 4×4) matrices. Proving this statement is not particularly difficult (see the formula described in the previous chapter on the parametric equation of a line). However, proving theorem 2.17 is much more complicated; see for example [Keré66] or [Fisc85] for the detailed proof.

Clearly, theorem 2.17 has the same importance for computer graphics as the existence of homogeneous coordinates, and for the same reasons: it becomes feasible to manage the collinearities numerically.

If the homogeneous coordinate system is generated out of a Cartesian one (following the method described in a previous chapter), the matrix representation gives also an easy way to decide whether a transformation is affine or not. Namely:

Theorem 2.18. If T is a non-singular projective transformation of $I\!PE^2$ to $I\!PE^2$ and $\overline{T}$ is its matrix representation, T is affine if and only if $\overline{T}$ is of the form:

$$\begin{bmatrix} t_{1,1} & t_{1,2} & t_{1,3} \\ t_{2,1} & t_{2,2} & t_{2,3} \\ 0 & 0 & \lambda \end{bmatrix} \tag{2.23}$$

where λ is a non-zero real number.

In case of a transformation in the projective space, the corresponding form is

$$\begin{bmatrix} t_{1,1} & t_{1,2} & t_{1,3} & t_{1,4} \\ t_{2,1} & t_{2,2} & t_{2,3} & t_{2,4} \\ t_{3,1} & t_{3,2} & t_{3,3} & t_{3,4} \\ 0 & 0 & 0 & \lambda \end{bmatrix} \tag{2.24}$$

Theorem 2.18 is very easy to prove: spatial ideal points are uniquely characterised by the fact that their last coordinate value is zero, in other words, they are of the form $[(\alpha,\beta,\gamma,0)]$ where α,β and γ are arbitrary real numbers with at least one of them being non–zero. The fact that the transformation T is affine means therefore that

$$t_{4,1}\alpha + t_{4,2}\beta + t_{4,3}\gamma + t_{4,4}0 = 0 \tag{2.25}$$

for all possible non all-zero choices of α, β and γ. This means that $t_{4,1} = t_{4,2} = t_{4,3} = 0$ should hold; the value of $t_{4,4}$ must however be non-zero, to ensure non-singularity. ∎

The reader may recognise the so–called *segment transformations* defined in GKS, GKS-3D, CGI etc. ([ISO85], [ISO88], [ISO88a]). However, the *modelling transformation* of PHIGS, PHIGS PLUS or PEX ([ISO89], [ISO89a], [ISO88b]) are not necessarily affine ones; indeed, the specification in these documents allows the user to give a general 4×4 matrix, without specifying any special features for the last row of it. This fact has severe algorithmic consequences on the so called *modelling clip* feature of these latter systems; more about that later. It is, however, strange that the GKSM specification in both the official GKS and the GKS-3D documents ([ISO85], [ISO88]) permit full 3×3 (resp. 4×4) matrices for segment transformations; this is clearly a mistake in the specification, as it would require the ability to handle a full, not necessarily affine transformation which is in contradiction with the rest of the specifications.

The different affine transformations used in computer graphics (rotations, scalings, shearings and translations) and their matrix representations are well described in a number of computer graphics textbooks (see all the already cited references) and it makes no particular sense to repeat these descriptions here[t]. In fact, the uniformity in description offered by the use of matrices was one of the reasons why homogeneous coordinates have become widespread in computer graphics even for 2D problems; it is a pity that even the newest textbooks on 3D graphics (like [Watt89]) do not explain why their use is mandatory when using projective transformations.

2.7.2. Viewing and Modelling Transformation

The main target of 3D systems is, after all, to render three dimensional objects on a two dimensional surface, which is the display screen or a plotter output. For that purpose, each such system has an internal mechanism which is usually called *viewing*. The most widespread approach to viewing is what is called the *synthetic camera model* in computer graphics literature and which is shown on figure 2.9. The idea is to project objects in a three dimensional frustrum onto the view plane; the frustrum (called the *view volume*) also serves as a clipping volume in space.

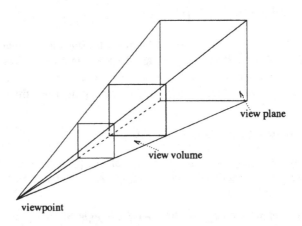

Figure 2.9.

The situation shown in figure 2.9 is usually referred to as "central projection" as opposed to the case where the the viewpoint is an ideal point, in which case the projection becomes a "parallel projection". In projective geometry terms, there is no difference between these two versions although, of course, the computational demands of a central projection are much greater than of a parallel one.

---

[t]Care should be taken, however, when using the different textbooks. Following the differences in conventions regarding operator–argument versus argument–operator notations in algebraic formulae, some of the textbooks might use vector–matrix multiplication rather than the convention adopted here. In such cases, the described matrices should be transposed to generate the appropriate version.

Early 3D graphics systems have effectively performed the projections as shown in the figure. This involved, however, two disagreeable consequences:

- clipping against the frustrum involves clipping against arbitrary planes in $\mathbb{R}^3$; this step is algorithmically demanding and

- performing the Hidden Line and/or Hidden Surface Removal in such a case is quite complicated as well.

As a result, in all newer systems (as well as in all 3D Graphics Standards or proposed Standards) the projection is done by first performing a special projective transformation in space (that is in $\mathbb{PE}^3$) which transforms the view volume onto a sub–cube of the unit cube and which transforms the viewpoint onto the ideal point of the z axis of $\mathbb{R}^3$ (see figure 2.10). Having performed this transformation the projection itself becomes simply the projection onto the x–y plane, the clipping against a view volume is reduced to a clipping against planes parallel to the base axes and, finally, the Hidden Line/Hidden Surface removal can be performed by applying appropriate algorithms which use exclusively the relative magnitude of the z coordinate values of points (see for example [Mudu86] for an overview of such algorithms).

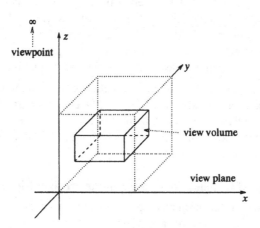

Figure 2.10.

The theorem about the existence and the uniqueness of a projective transformation (theorem 2.16) ensures that a mapping of the view volume onto the sub–cube of the unit cube uniquely exists. In general, this mapping will be non–affine (more precisely, the affinity of the transformation is equivalent to a parallel projection). There are several methods for the calculation of this transformation (also called *view transformation*); textbooks or tutorials like [Fole84], [Watt90], [Herm91], [Fole90] or others all describe the necessary steps and K. Singleton has also given a good and detailed overview of the GKS–3D/PHIGS cases ([Sing86]).

In this latter work, a description of the usual classification of the projections themselves is also given (oblique, isometric, 2–point perspective etc.); all these different mappings produce different visual effects on the screen, but their mathematical backgrounds are all the same. It is unnecessary to repeat all these formulae here; the reader is referred to the cited works (or alternative ones).

It has to be stressed that the ISO 3D documents do *not* state that the view transformation should be of the format described above. Not all projective transformations transform such a view frustrum onto a regular cube; there might be more complicated ones as well (eg the planes describing the view volume might not be parallel to the view plane). The user of a GKS–3D/PHIGS system has the possibility to specify any kind of 4×4 matrix as a view transformation and the system should be able to deal with it. The process described above is just offered as a utility (via the so–called utility functions of the specifications).

In PHIGS (and PHIGS PLUS), the viewing pipeline of the system includes yet another transformation which is called the *modelling transformation*. This transformation is primarily aimed at the use of convenient local coordinate systems for describing objects in 3D; by specifying the modelling transformation the appropriate coordinate transformation can be performed by the graphics system itself (in this sense, this transformation is the generalisation of the so called normalisation transformation of the GKS and GKS–3D specifications, [ISO85], [ISO88a]). However, to achieve some special effects (like the image of a projection modelled by three dimensional objects) this transformation is to be specified by a full 4×4 matrix, allowing therefore a general, not necessarily affine transformation.

The view transformation and the modelling transformation are the two, not necessarily affine, transformations arising in 3D computer graphics systems. In what follows, no special attention will be paid to their usual use in practice and their format; as described above, the 3D systems at hand must be able to handle these transformations in all their generality.

2.7.3. Description of Collinearities Based on the Straight Model

The straight model of the projective plane and/or space also permits the visual representation of the effects of a projective transformation. This is important because just as the straight model creates a strong link between $I\!\!P\!E^2$ and $I\!\!R^3$ (respectively between $I\!\!P\!E^3$ and $I\!\!R^4$), the visualisation to be described creates a link between projective transformation and the linear transformation of $I\!\!R^3$ and $I\!\!R^4$ respectively. This is of no particular interest in traditional projective geometry and it is therefore never described in the previously cited projective geometry textbooks; however, for the purposes of computer graphics where the linearity of some traditional problems has importance, its use has proven to be extremely useful (see [Herm87], [Herm88], [Herm91], [Hübl90]).

As in the case of the projective points and lines of $I\!\!P\!E^2$, projective transformations described acting on $I\!\!P\!E^2$ can be visualised easily. It is worth recalling (see also figure 2.11) that in the straight model $I\!\!P\!E^2$ is identified with the points in $I\!\!R^3$ on

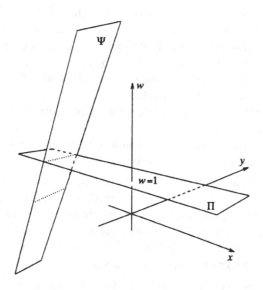

Figure 2.11.

the plane $w = 1$ (denoted by Π on figure 2.11). A projective transformation T is represented by a matrix $\overline{T}$. This matrix $\overline{T}$ also generates a traditional linear transformation T' in $I\!\!R^3$ using the matrix-vector multiplication. This transformation will map Π onto another plane of $I\!\!R^3$ which has, in general, an arbitrary position in space. This plane is denoted by Ψ on the figure. (In fact, Ψ can also be considered as an alternative straight model of $I\!\!PE^2$). To get back to the more usual $w = 1$ model, Ψ has to be projected centrally (through the origin) back onto Π; this is, in fact, the so–called projective division.

What can be said therefore, in view of figure 2.11, is as follows. A transformation T of $I\!\!PE^2$ can be modelled by a *two–stage process*. First, Π is mapped by the *linear* transformation T' (of $I\!\!R^3$) onto Ψ and, secondly, the resulting plane Ψ is projected back onto Π by a central projection through the origin. The first step might be called the *linear part* of the transformation while the second step is traditionally denoted as the projective division. This simple fact has far–reaching practical consequences; in fact, the whole of the next chapter will be based, in some sense, on this observation.

Clearly, T is affine if and only if Ψ is parallel to Π (on figure 2.11). Furthermore, the usual description of the affine transformation (that is where $T_{4,4} = 1$) results in $\Pi = \Psi$.

In the case of $I\!\!PE^3$, Π and Ψ are three dimensional subspaces of $I\!\!R^4$, also called hyperspaces. With this difference, the whole description remains valid for $I\!\!PE^3$ as well.

2.8. Division Ratio and Cross Ratio

As stated before, the notion of line segments has no meaning in projective geometry any more. Likewise, the distance between two points also becomes an unclear notion; indeed, the distance is usually defined in terms of the generated line segment. Furthermore, even if the distance could be defined at least in restricted cases (eg by excluding ideal points), the projective transformation does not retain these values. It is to find some kind of a stable numerical value that the notion of double ratio has been introduced in projective geometry; in a number of cases, its use is necessary to prove some of the important statements of the theory. Furthermore, as will be presented later, the double ratio can also be a very useful tool for the purposes of computer graphics, that is the reason why it is presented here.

The division ratio and the double ratio will be defined in terms of affine and ideal points, that is making use of the construction which has led to a projective plane and/or projective space. A mathematically "pure" definition (that is based on the axiomatic system only) would also be possible, but it would be more abstract and more complicated as well (see eg [Coxe74]). For the purposes of computer graphics, the more "pragmatic" approach is acceptable.

At first, the notion of *division ratio* has to be defined as follows.

Definition 2.7. Let A, B and C be three different collinear affine points on IPE^2 or IPE^3. The *division ratio* of the points A, B and C (denoted by (ABC)) is defined to be the following (non-zero) real number:

$$(ABC) = \frac{\overline{AC}}{\overline{CB}} \tag{2.26}$$

where $\overline{AC}$ and $\overline{CB}$ mean the *directed* Euclidean distances of the two points (that is $\overline{AC} = -\overline{CA}$).

If the point C tends to infinity, the limit of the corresponding division ratio (with the points A and B remaining fixed) will be -1; consequently, it seems to be feasible to extend (2.26) to the case when the point C is an ideal point, namely let

$$(ABC) = -1 \tag{2.27}$$

in this case.

With the help of the division ratio the *double ratio* (also called sometimes the *cross ratio*) of four collinear points may be defined as follows.

Definition 2.8. Let A, B, C and D be four different collinear points of IPE^2 or IPE^3 not all four being ideal. The *double ratio* of the four points (denoted by $(ABCD)$) is the real number defined by:

$$(ABCD) = \frac{(ABC)}{(ABD)} \tag{2.28}$$

Clearly, when all four points are affine (i.e. none of them is ideal), the double ratio may also be expressed with the help of directed distances, namely:

$$(ABCD) = \frac{\overline{AC}\,\overline{DB}}{\overline{CB}\,\overline{AD}} \tag{2.29}$$

The double ratio has a number of remarkable properties. Some of them will be cited here, which are necessary for later purposes; the corresponding proofs may be found for example in Penna and Patterson ([Penn86]), in Fischer ([Fisc85]) or in any other standard textbook on projective geometry[†].

First of all, the double ratio is a one–to–one mapping of the points of the line and the (non-zero) real numbers. Namely, the following is true:

Theorem 2.19. Given three different collinear points on the projective plane, denoted by A, B and C, if x is an arbitrary non-zero real number, then there exists one and only one point D on the line determined by A, B and C, for which the following equation holds:

$$(ABCD) = x \tag{2.30}$$

The second property has a particular importance for computer graphics (and, in fact, it is one of the most important results in projective geometry). The description of this property requires first a definition:

Definition 2.9. The projective invariance of the double ratio is defined as follows. If A, B, C and D are four different arbitrarily chosen collinear points of the plane (or space) and T is an arbitrary projective transformation, then the following property should be valid:

$$(ABCD) = (T(A)T(B)T(C)T(D)) \tag{2.31}$$

The projective invariance of the division ratio means that for each three points A,B and C the relation

$$(ABC) = (T(A)T(B)T(C)) \tag{2.32}$$

holds.

The affine invariance of the double ratio and the division ratio can be defined similarly; in this case T should be affine.

[†]One should be careful again, however, when consulting the literature; in some cases, following different local traditions, the definition of the double ratio may slightly differ from the one given here (e.g. by an additive constant, the order of the directed distances in the formulae etc.).

With this definition at hand, the following theorem holds:

Theorem 2.20. The double ratio is projective invariant, the division ratio is affine invariant.

See for example [Fisc85] or [Penn86] for a detailed proof.

Theorem 2.20 also shows that the value of the double ratio is independent of the construction of IPE^2/IPE^3. Although this has not been explicitly stated up to now, the Euclidean analogy works here perfectly, and the change of homogeneous coordinate system on IPE^2 or IPE^3 can be described by a matrix–vector multiplication, using, of course, a homogeneous matrix. In other words, this is analogous to the use of a projective transformation which, according to theorem 2.20, leaves the value of the double ratio unchanged.

It should be remarked that a stronger statement regarding invariance of the division ratio is not true. That is, the the division ratio is *not* projective invariant.

Penna and Patterson in [Penn86] give some methods to calculate the exact value of the double ratio for four given collinear points in IPE^2. Without going into details, the idea is to project the points onto the main coordinate axes (which can be done, in fact, by replacing one of the coordinate values by 0) and calculate the double ratio of the resulting four points. Taking into account that the projection keeps the value of the double ratio, this is clearly a valid approach.

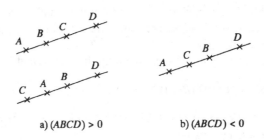

a) $(ABCD) > 0$ b) $(ABCD) < 0$

Figure 2.12.

In some cases, however, not the exact value but only the *sign* of the double ratio is of interest. Indeed, it is easy to prove (using formula (2.29), see also figure 2.12) that if for the points A, B, C and D the value of $(ABCD)$ is negative, the (affine) line segments AB and CD "cut" (overlap) one another while that is not the case if the value of $(ABCD)$ is positive (see figure 2.12). The importance of this fact is that as an arbitrary projective transformation keeps the value of the double ratio, this "segment cutting" property is invariant for projective transformations.

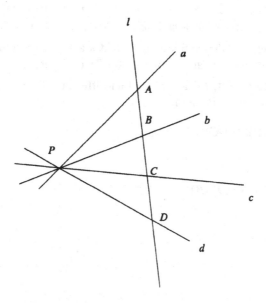

Figure 2.13.

Theorem 2.20 permits the assignment of a double ratio to four lines of $I\!P\!E^2$ which have one common intersection point (such a configuration will be referred to as a "bunch of lines"). Figure 2.13 shows how this can be done. If a, b, c and d are the four lines intersecting at P, let us take any line l *not* containing P. If $l \wedge a = A$, $l \wedge b = B$ etc., the following definition can be given:

$$(abcd) = (ABCD) \tag{2.33}$$

The fact that this assignment can be done is by itself is not very surprising; indeed, a bunch of lines (that is four lines intersecting at one point) is the dual form of four points lying on the same line. The definition is not dependent upon the choice of the line l: if another line, say l' were taken, the central projection with P as a centre would map l onto l', mapping the corresponding intersection points onto one another; consequently, according to theorem 2.20, the definition in (2.33) is indeed correct. This also means that the "segment cutting" property has also its counterpart for such a configuration, although it should rather be called "domain cutting" in this case.

Formula (2.33) also fills a "hole" in the definition of the double ratio. If four collinear points A,B,C and D are all ideal, there has been no definition given up to now for the value of $(ABCD)$. However, four ideal points generate four lines by choosing an arbitrary point P on $I\!P\!E^2$; formula (2.33) gives then the value of $(ABCD)$. Using the theorem on the existence of a projective transformation (theorem 2.16), if a different point Q is chosen instead of P, there exists a projective transformation which would transform the corresponding lines onto one another; as the double ratio is projective invariant for lines also, this definition holds.

The invariance of the double ratio provides a method to prove a number of so-called permutation formulae on double ratio. Two of them are as follows:

Theorem 2.21. If A,B,C and D are four different collinear points, then the following equalities hold:

$$(ABCD)(ABDC) = 1 \tag{2.34}$$

and

$$(ABCD) = (CDAB) \tag{2.35}$$

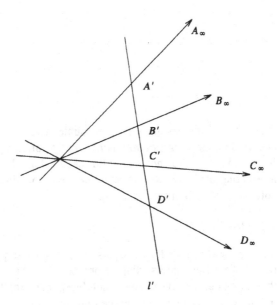

Figure 2.14.

The proof for these formulae is very simple: it is enough to prove the case where none of the points A,B,C or D is ideal. If this is not the case, it is possible to choose four appropriate affine points A',B',C' and D' and a projective transformation (actually a central projection) which would map one set of points onto the other. Figure 2.14 shows the case when all four points are ideal (and projected onto the line l') and figure 2.15 is the case when only one of the points is ideal (and projected onto the line l' again). The points being collinear, there are no other alternatives. If the equations are true for the points A', B', C' and D', they are also true for A, B, C and

D because of the invariance of the double ratio. Finally, to prove that the equations are true for the purely affine case is simply a matter of algebraic exercise, based on the definition of the double ratio. ∎

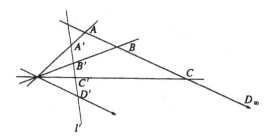

Figure 2.15.

For IPE^3, a double ratio can be assigned to each four *coplanar* lines intersecting in one point. Furthermore, if the planes Π, Ψ, Φ and Θ are such that they intersect in one common line, the value of the double ratio $(\Pi\Psi\Phi\Theta)$ can be defined as well, using formula (2.33); this also means that the "domain cutting" property is also valid in this case.

In [Kram89] G. Krammer introduced the notion of *conic sectors*, which is an interesting application of the double ratio. In what follows, conic sectors will be defined for IPE^2; in case of IPE^3, planes should be taken instead of lines to arrive at the same definitions.

If two lines, say l and n, are given on IPE^2, these lines will cut IPE^2 into two disjoint subareas. The definition is as follows.

Definition 2.10. If $P,Q \in IPE^2$ are given, let X (resp Y) denote the intersection points $(P \vee Q) \wedge l$ (resp. $(P \vee Q) \wedge n$). In case $X = Y$, the points P and Q are said to be in the same conic sector. If $X \neq Y$, P and Q are said to be in the same conic sector if and only if the following inequality holds:

$$(PQXY) > 0 \tag{2.36}$$

See also figure 2.16. There might be some special cases for conic sectors. If l and n are parallel in the Euclidean sense (that is their intersection point is an ideal point), the corresponding conic sector is shown in figure 2.17a). If, finally, l is affine but n is the ideal line, the two conic sectors are the two half-planes! (see figure 2.17b).

The interesting thing about conic sectors is the fact that conic sectors are transformed into conic sectors by projective transformations. Indeed, projective transformations map intersection points onto intersection points; that is, if T is the transformation in use then, using the notation of figure 2.16:

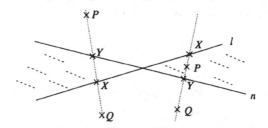

Figure 2.16.

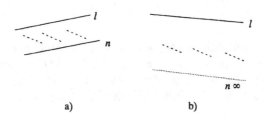

a)   b)

Figure 2.17.

$$(T(P) \vee T(Q)) \wedge T(l) = T(X)$$
$$(T(P) \vee T(Q)) \wedge T(n) = T(Y)$$

(2.37)

furthermore, the projective transformation also keeps the double ratio, that is

$$(PQXY) = (T(P)T(Q)T(X)T(Y))$$

(2.38)

in other words, the sign of the double ratio will also remain unchanged.

In Euclidean geometry, convex polygons or convex polyhedra can be described by the intersection of a finite number of half planes/spaces. Conic sectors may be considered as the generalisation for the projective case of half-spaces (or half-planes); in other words, one can speak about projective polygons as the intersection of a finite number of conic sectors. The importance of this approach will become clear later.

2.9. Projective Theory of Conics

2.9.1. Introduction

Besides line segments and polygons, conics also occur frequently in computer graphics. They are used in geometric design, they are frequently implemented as GDPs (Generalized Drawing Primitives) in various graphics packages and some are included in the basic set of output primitives of ISO documents (e.g. CGI, [ISO88a]). Among the three main classes of conics, namely ellipses, parabolae and hyperbolae, the use of ellipses (first of all circles) is the most widespread. Circles and circular arcs are used in business graphics for charts, in mechanical engineering for rounded corners, for holes etc. Circular arcs may also be used to interpolate curves (see for example Sabin in [Sabi77]). Although the role of parabolae and hyperbolae is not so important, they cannot be ignored either. On the one hand they do appear in practical applications (for example there are proposals to use parabolic arcs for curve approximation like the so-called double-quadratic curves in Várady in [Vára84] or [Vára85]) but, principally, these curves appear automatically when distorting an ellipse with a projective mapping.

Projective geometry gives a unified framework for handling all kinds of conics. This might sound surprising at first glance, as projective geometry is considered to be a theory primarily concerned with lines and their behaviour. However, if one thinks of the well-known fact that the conics appear as the planar intersections of cones, which is very much like the figure of a central projection, it becomes more plausible that projective geometry has this descriptive power for conics.

What is the real problem as far as computer graphics is concerned, in handling these curves? Mathematically, (planar) conics are described by a second order polynomial of the form:

$$a_{1,1}x_1^2 + a_{2,2}x_2^2 + 2a_{1,2}x_1x_2 + 2a_{1,3}x_1 + 2a_{2,3}x_2 + a_{3,3} = 0 \tag{2.39}$$

While this formula is appropriate to perform all calculations which are necessary in a modelling system (see for example Fraux and Pratt [Faux79]) it is inadequate to *draw* the corresponding conic. Indeed, practically all graphics devices available today are designed to render (in hardware/firmware) line segments; in other words, the "ideal" mathematical curve must be approximated by an appropriate polyline or polygon. To achieve a reasonable appearance, this approximation must be quite dense; for example the number of approximation points to render a circle properly must be at least 100, but an approximation with 360 points (that is one point for each degree) may also be necessary.

It is not an easy task to generate these points properly. Appropriate approximation formulae or equations are necessary; some examples will be presented later. Some of these formulae (especially those describing ellipses) are already known to the graphics community, whereas some others are relatively unknown. Furthermore, having these formulae at hand does not solve all problems. An implementor has to give an answer to the following question: where in the graphics output pipeline is the approximation effectively performed?

The approach usually chosen is to approximate the curve with a polygon/polyline before performing a transformation in the pipeline. Most of the formulae described in different textbooks and used in practice are not projective invariant, that is the data generating these formulae change their geometrical nature when applying a projective transformation. Therefore, the curves are approximated beforehand, the resulting polygons/polylines are transformed and rendered following already well established methods.

There are, however, some problems with this approach. First of all, there is a loss in speed and storage. As mentioned already, the number of generated points tends to be relatively large; all these points have to be transformed, that is a matrix–vector multiplication has to be applied and, in the case of a projective and non–affine transformation, an additional projective division must also be performed. By applying some alternative methods presented later (primarily in chapter 5), a speed improvement of at least 25% can be achieved. This figure might not seem very impressive at a first glance but one should never forget that computer graphics is (ideally) interested in real–time effects where a 20%-25% improvement might be of real significance.

A widespread approach to overcome this difficulty is to use the so–called rational B–splines to describe conics. Second order rational B–splines can describe any kind of conic and this is done in a more or less projective–invariant manner (see eg [Faux79], [Till83], [Pieg87], [Fari88]). Beside the fact that B–Splines are computationally expensive (see all the calculation formulae in [Bart85] or [Bart87]), this approach leads still to another common problem: the quality of the approximation.

Speed is not the only issue (and having all these super–fast computers invading the market, this argument might be less and less important). However, when approximating for example a circle with 360 points, one gets a fairly regular geometrical ordering of the points which, if displayed directly, will produce an acceptably smooth shape. However, if a transformation is applied against this set of points, this "regularity" will be lost. Some of the line segments will become much longer than others; in these areas the resulting polyline will have a "jagged" effect whereas on some other parts of the curve the density of the points will be unnecessarily high. It is very difficult to keep track of these distortions which may be, in the case of a more complicated projective transformation, very noticeable. The only way of reducing this effect is to postpone the approximation step as "far" as possible and to produce the resulting polyline after the transformations instead of prior to it.

The real difficulty with this approach is the fact that a non–affine projective transformation will "destroy" a number of geometrical characteristics of the points. As an example remember that the centre of an ellipse might not be the centre any more; furthermore, the image of an ellipse is not even an ellipse in some cases; it may become a hyperbola or a parabola. Consequently, a thorough investigation of the nature of conics is necessary, using the tools of projective geometry. This is why this part of the theory has also been included in the present study.

2.9.2. General Theory of Conics

2.9.2.1. The 2D Case

As usual, planar conics will be treated first; some of the ideas will then be generalised for space as well.

In a Euclidean environment, a conic is described by the equation (2.39). By defining the symmetric matrix $A = (a_{i,j})_{i,j=1}^3$ and by using homogeneous coordinates instead of Euclidean ones (with the usual identification mechanism), the equation has its counterpart for projective environments as well, namely:

$$\sum_{i,j=1}^{3} a_{i,j} x_i x_j = 0 \tag{2.40}$$

Finally, formula (2.40) can be abbreviated by the so called bilinear form, that is:

$$xAx = 0 \tag{2.41}$$

The notation of formula (2.41) will be used throughout the whole chapter. In the whole section A will be considered to be a non-singular matrix, that is $det(A) \neq 0$ (some of the theorems presented later are not valid for singular cases; on the other hand, the corresponding "curves" in this case are lines, points or just the empty set).

In fact, (2.41) can be used in a somewhat more general way to define the notion of *conjugate points*. This definition is as follows:

Definition 2.11. The points $x, y \in IPR^3$ on the projective plane are said to be conjugate points with respect to the conic represented by the symmetric matrix A if and only if the following equation holds:

$$xAy = 0 \tag{2.42}$$

(A being symmetric, $xAy = yA^Tx = yAx$ holds). We could also say that the points of the curve may be characterised by the fact that they are auto-conjugate.

The notion of conjugate points has a number of nice properties. Indeed, the following facts are true (their proofs may be deduced from the definitions or they may be found in the textbooks cited above):

Theorem 2.22. If $x \in IPR^3$ is a fixed point, the set of all points $y \in IPR^3$ which are conjugate to x with respect to a conic represented by the symmetric matrix A form a line of the projective plane. This line is called the *polar* of x; it may be represented by the homogeneous vector Ax.

Theorem 2.23. If l is a line in the projective plane, then there is one and only one point whose polar with respect to a conic represented by

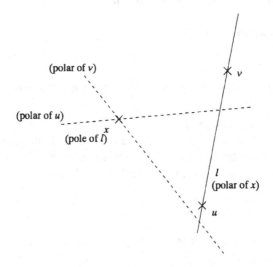

Figure 2.18.

the symmetric matrix A is l; this point is called the *pole* of l. The pole may also be characterised as follows: it is the (unique) intersection point of all the polars generated by the points on l (see also figure 2.18).

Theorem 2.24. If $x \in I\!PR^3$ is the homogeneous vector of a point on the projective plane, then x belongs to its own polar if and only if x is a point of the conic itself. In this case, the polar of x will be tangential to the conic at the point x and the homogeneous representation of this tangential line is Ax.

Definition 2.12. The pole of the ideal line is called the *centre* of the curve; for ellipses and hyperbolae, this coincides with the traditional, Euclidean definition of the centre of these curves.

Two lines l_1 and l_2 are said to form a *conjugate pair of lines* if the pole of l_1 belongs to l_2 and, conversely, the pole of l_2 belongs to l_1. One may speak of a pair of *conjugate chords* as well as of a pair of *conjugate diameters*, denoting a pair of conjugate lines which are chords (resp. diameters) of the conic (diameter is a chord containing the centre).

All these definitions are, unfortunately, rather abstract and a certain time is needed to get used to them and to get an intuitive feeling as far as their geometrical meaning is concerned. Figure 2.19 shows an example which might help in using these definitions. The line l has two intersection points with the conic, P and Q. The polars of these points are the two tangents l_1 and l_2 respectively. In view of what has been said before, the intersection point of these lines, that is R, is the pole of the

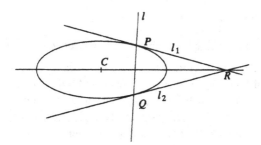

Figure 2.19.

line l. This procedure is the usual way of generating the pole of a line, provided the line has two intersection points with the curve (which is not always the case). It can also be remarked that if the centre of the curve is denoted by C (like in the figure), the line $R \vee C$ will intersect the line segment PQ in its middle point. Also, the pole of the lin $R \vee C$ will be on l, that is a pair of conjugate lines have been created (see for example [Keré66] for a proof of these features).

The importance of these definitions becomes clearer when the behaviour of conics in relationship to projective transformations is examined. If the matrix of the transformation is denoted by T, and if a conic is represented by the symmetric matrix A, then for all $x, y \in I\!\!PR^3$:

$$xAy = x^T(Ay) = = x^T(T^{-1}T)^T(AT^{-1}Ty) = \qquad (2.43)$$
$$x^T T^T(T^{-1})^T(AT^{-1}Ty) = (Tx)^T((T^{-1})^T AT^{-1})(Ty)$$

Now, if the notation

$$T(A) = (T^{-1})^T A(T^{-1}) \qquad (2.44)$$

is applied, then (2.44) can be simplified to $(Tx)T(A)(Ty)$. In other words, the image of a conic under the effect of a projective transformation remains a conic and, furthermore, formula (2.44) gives an easy way to calculate the matrix of the image. Also, the property of conjugation is projective invariant. The pole–polar relationship also remains valid across the transformation. However, the image of a centre is *not* necessarily a centre: although it is true that the image of the centre will still be the pole of the image of the ideal line, it is not necessarily the case that the image of the ideal line will still be the ideal line (it is however affine invariant).

The regularity of the matrix A has an interesting consequence, which is as follows.

Theorem 2.25. If $p, q \in I\!\!PR^3$, $p, q \neq 0$ are such that

$$pAp = qAq = pAq = 0$$

then $p = q$ (where the equality is meant in homogeneous sense).

Let $u = Ap$ and $v = Aq$. The vectors u and v are usual three dimensional vectors, that is $u, v \in I\!\!R^3$. None of them is the zero vector, as $det(A) \neq 0$. If u and v are equal in the homogeneous sense, then $p = q$ follows (again in the homogeneous sense). If this is not the case, then, from the premises of the theorem, p as a regular three dimensional vector is perpendicular to both u and v. In other words, p is perpendicular to the plane uvv. However, the same is true for q and this is possible only if there exists a $\lambda \in I\!\!R$ such that $p = \lambda q$. ∎

The mutual relationship of a conic and a line is of particular interest. Namely:

Theorem 2.26. The number of intersections of a line and a conic may be 0, 1 or 2.

Although this fact is well-known in projective geometry, it is worth presenting the proof of this theorem here as well. The reason is that the proof gives an effective way of calculating the (possible) intersection points and this is a useful feature in what follows. It will also be necessary to make use of the result of theorem 2.25.

The line to intersect with can be described by (see (2.18)):

$$p \vee q = \{ \lambda p + \mu q \} \tag{2.45}$$

$$\lambda \neq 0 \text{ or } \mu \neq 0$$

where p and q are two points on the line. Two values for λ and μ are searched for which

$$(\lambda p + \mu q) A (\lambda p + \mu q) = 0 \tag{2.46}$$

holds. In fact, because of the homogeneous nature of the formula, one is not interested in the exact values of λ and μ but only in their relative ratio λ/μ. Equation (2.46) can be also rewritten by:

$$pAp\lambda^2 + 2pAq\lambda\mu + qAq\mu^2 = 0 \tag{2.47}$$

First of all, pAp and qAq cannot be both zero. Indeed, this would mean

$$2pAq\lambda\mu = 0 \tag{2.48}$$

for all possible choices of λ and μ, that is $pAq = 0$ would also hold; however, according to theorem 2.25, this is not possible (p and q are considered to be different points of $I\!\!PE^2$). If $pAp \neq 0$, $\mu \neq 0$ may be considered; if this were not the case, then $\lambda = 0$ would also hold, which is impossible. Similarly, if $qAq \neq 0$ then $\lambda \neq 0$. Let us consider the first case; that is the equation can be divided by μ^2 to get

$$pAp(\lambda/\mu)^2 + 2pAq(\lambda/\mu) + qAq = 0 \tag{2.49}$$

clearly, this equation (in λ/μ) may have 0, 1 or 2 solutions; by solving it one also gets an explicit value for the (possible) intersection point(s). ∎

The relationship between lines and conics has a very important consequence, as the theorem can be applied to a special case to get a simple means of classification for conics. Namely:

Theorem 2.27. The number of ideal points belonging to a conic may be 0, 1 or 2. (The set of ideal points being the ideal line, this is just the special case of the previous statement). If this number is 0, the curve is an ellipse (or a circle); if it is 1, the curve is a parabola with the axis determining its ideal point; and if it is 2, the curve is a hyperbola, with the two asymptotes determining the two ideal points.

Theorem 2.27 (and the previously cited features) are of particular importance for computer graphics. Some of the consequences are:

- The tangent of a curve remains a tangent and a chord remains a chord after transformation: the intersection points of lines and curves are auto-conjugate points and the conjugation is a projective invariant property.

- Each class of conics is *affine invariant*. In other words, the affine image of an ellipse will be an ellipse, the affine image of a parabola will be a parabola etc. In the case of parabolae for example, the ideal line is tangential to the curve; the image of a tangent being still a tangent and the image of the ideal line being still the ideal line, the image curve has only one ideal point. In other words, the image of the curve has still one ideal point only, which means, according to theorem 2.27, that the image is a parabola. The same reasoning holds for ellipses and hyperbolae as well.

- By using straightforward and simple calculations it is easy to decide from the matrix of a curve which class the curve belongs to: the way theorem 2.26 was proven gives also a way to calculate the intersection points (if any), and also to calculate their number. All what is required, is to use the homogeneous coordinate values of two ideal points (see 2.6.1 for IPE^2 and 2.6.2 for IPE^3).

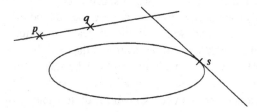

Figure 2.20.

Let this section be concluded by yet another calculation formula for planar conics which will be useful later. The task is as follows: if $p, q \in IPR^3$ are two points, A is a symmetric matrix representing a conic and, furthermore, $s \in IPR^3$ is a point on the conic, compute the intersection of the tangent at s and pvq (see figure 2.20). This could be done reusing already known formulae but an alternative (and computationally more attractive) method is as follows. Once again, appropriate λ and μ numbers are to be found so that:

$$(\lambda p + \mu q) A s = 0 \qquad (2.50)$$

taking into account that As gives the homogeneous representation of the tangential line at s. That is:

$$\lambda p A s + \mu q A s = 0 \qquad (2.51)$$

Again, if $pAs \neq 0$ then $\mu \neq 0$, that is we can divide; the result is:

$$\lambda/\mu = -qAs/pAs \qquad (2.52)$$

2.9.2.2. The 3D Case

The notion of conics may be generalised for projective spaces as well; the only difference is that the symmetric matrix in use should be 4×4 instead of 3×3. These conics are the so–called quadratic surfaces (hyperboloids, paraboloids, hyperbolic paraboloids etc.). Their classification is much more complicated than in the case of planar curves; however, they form again a class of surfaces which is invariant to projective transformations (the way theorem (2.44) has been deduced was independent of dimensions). Many of these quadratic surfaces are rarely used directly in computer graphics, except some of the symmetric rotation surfaces. In such cases, the rational B–spline formulation for these surfaces is the widely accepted approach (see again [Till83], [Pieg87], [Fari88] and also [Klem89]).

As quadratic surfaces appear in very special cases only, no further investigation will be presented here. Instead, this section will concentrate on what will happen to planar quadratic curves in a 3D environment; taking into account their usefulness in practice, it is worth examining this special case more thoroughly.

One way of handling planar conics in space is to find a vector representation of a conic curve, that is an equation which describes the points of the curve as a function of some vectors and some additional real parameters. Such a formula would be useful if it were at least affine invariant, that is if the transformation of the vectors of the equation were enough to describe the transformed curve. Such an affine invariant formula can be found for all three classes of the curves; while the formula describing an ellipse has been known for quite a long time, the corresponding formulae for parabolae and hyperbolae had to be reconstructed from different mathematical bits and pieces (these formulae were never of a real interest to mathematicians, that is why they are not usually presented anywhere). This has been done in [Herm89a] and will also be presented in a later chapter.

To perform some computations, however, a more complicated approach is also necessary which is as follows. The intersection of a plane and a quadratic surface leads to a planar conic on the plane. If the plane happens to be the plane $x_3 = 0$, this can be easily seen by just putting a 0 to all relevant places of the equation of the surface; the result is a second order equation for the remaining coordinate values. If the plane is of a general position, it can always be transformed into the plane $x_3 = 0$ by using an orthogonal transformation. These intersection curves are from now on the main focus of interest. Also, it is easy to associate a quadratic surface with a planar conic: one has to construct a *generalised cylinder* (it might also be called a sweep surface). This means that the curve should be moved along a line not contained by the plane of the conic (see figure 2.21). In the simplest case, when the conic lies in the $x-y$ plane, it is also very simple to give the equation of such a surface. If A is a 3×3 matrix then the matrix describing the corresponding surface may be:

$$A_c = \begin{bmatrix} a_{1,1} & a_{1,2} & 0 & a_{1,3} \\ a_{2,1} & a_{2,2} & 0 & a_{2,3} \\ 0 & 0 & 0 & 0 \\ a_{3,1} & a_{3,2} & 0 & a_{3,3} \end{bmatrix} \tag{2.53}$$

If the plane is not the $x-y$ plane, an affine transformation and 2.21 should be applied; concrete examples will be given in chapter 4 (see also [Herm89a]). Notice should be taken of the fact that the matrix A_c is singular; however, its rank is 3 (that is it contains a 3×3 non-singular submatrix). In fact, it can be shown that in case of 3D, if the matrix of a quadratic surface is singular but its rank is 3, it is either the matrix of a (generalised) cylinder or that of a cone (there is no difference in a projective sense between a cylinder and a cone: the cylinder is a cone whose focal point is ideal). For the proof of this theorem the interested reader should consult for example [Keré66].

Using (2.53), a three dimensional surface can be assigned to each planar conic. Using then (2.44), the image of this surface under the effect of a transformation can be described. As a next step some characteristic data of the *intersection of the surface and a plane* should be calculated; indeed, what a computer graphics system is really interested in is not the whole surface but only the planar cut of it. To use all the formulae of the previous section, the following are needed:

- If $p, q \in I\!PR^4$ are two points *in the plane* Π and A is a 4×4 symmetric matrix representing a quadratic surface, compute the number of intersection points and the eventual intersection points themselves of $p \vee q$ and the (planar section of) the surface.

The same calculations as the one presented for the two dimensional case can be adapted to 3D as well. Care should be taken, however, that in this case it is possible that the intersection is a line. Theorem 2.25 is not valid in $I\!PE^3$ any more (the arguments used to prove it were very much bound to the nature of $I\!E^3$); in other words, it is generally possible that a quadratic surface would contain a whole line (see

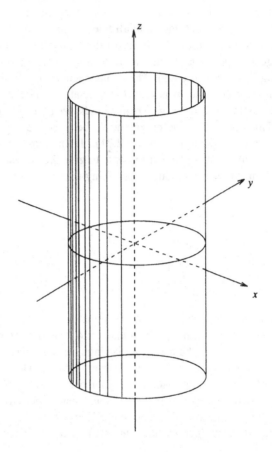

Figure 2.21.

figure 2.21 or, to choose a regular case, the well known saddle surface). But this situation is equivalent to the fact that $pAp = qAq = pAq = 0$, and this can be checked easily. In particular, if Π is the plane containing the original conic and A is the matrix of the transformed cylinder, by taking two ideal points of $A(\Pi)$, the exact classification *of the planar section* can be done.

- If $p, q \in I\!PR^4$ are two points in Π and A is a symmetric matrix representing a quadratic surface, compute the pole of $p \vee q$ according to the planar section of the surface.

Similarly to the two dimensional case Ap and Aq represent a (spatial) polar of p and q respectively. The difference is that the polar is now a plane instead of a line. However, calculating $(Ap) \wedge (Aq) \wedge \Pi$, leads to the two dimensional pole on Π (see section 2.6.2 for all the necessary formulae).

- If $p,q \in I\!P\!R^4$ are two points in Π, A is a symmetric matrix representing a quadratic surface and, furthermore, $s \in \Pi$ is a point on the planar intersection of the surface, compute the intersection of the tangent at s and pvq.

Essentially the same formulae can be used as in (2.52). Indeed, the tangential plane of A is given by As and the intersection of this plane with Π will give the tangent in Π.

- As presented later, these formulae (together with the ones listed in 2.6.2) will make it possible to generalise all two dimensional results into 3D. Examples for that will be presented later.

3. Practical Use of Four Dimensional Geometry

3.1. The W-Wraparound Problem

3.1.1. Introduction

Non–affine projective transformations map some affine (that is Euclidean) points onto ideal ones. This is the difference between non–affine and affine projective transformations. What will be the visible effect of this difference on the screen?

It is worth concentrating first on the simplest geometric primitive in use in computer graphics, that is a line segment. It has been shown, in section 2.7.3, that a projective transformation can be viewed as a two–stage process: first a linear transformation and secondly the projective division. In figure 3.1, which shows, as usual, the simpler two dimensional case, the line segment PQ on Π is transformed by the linear part of the transformation onto the line segment $P'Q'$ of Ψ and, in a second step, it is projected back onto the plane Π by the projective division. How-ever, as shown in the figure, something interesting occurs: the projection of the line segment $P'Q'$ is *not* the line segment $P''Q''$, but the complement of it, that is the union of the two half–lines determined by P'' and Q'' respectively.

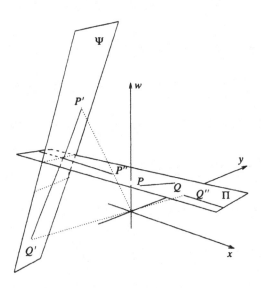

Figure 3.1.

The reason for this is that the line segment $P'Q'$ intersects the $w=0$ plane. This means, in projective geometric terms, that the image of the line segment PQ will contain an ideal point, that is it will *not* be the line segment $P''Q''$. Figure 3.2 shows this effect for the usual plane–to–plane central projection which was the starting point of the discussion. Finally, figure 3.3 shows a realistic situation with a schematic view of the synthetic camera model: here again, the line segment PQ is mapped onto two half–lines, generated by P'' and Q'' respectively. This last figure is also interesting because it shows in practice when the "danger" occurs: each line which intersects the vanishing plane of the projective transformation (which is, in the case of figure 3.3, the plane parallel to the view plane and containing the view reference point) will produce this strange result. In practical terms: if the model to be visualised is placed "around" the view reference point (eg an architectural CAD program which allows the view reference point to be put in the middle of a room), this situation will occur.

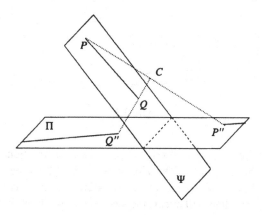

Figure 3.2.

The two half–lines generated by the projections have been given the name *external line segments* by Blinn and Newell in [Blin78]; this refers to the fact that the image of the line segment is, in projective terms, the set of projective points containing the two half–lines plus the ideal point of the line $P''vQ''$ which "glues" the two half–lines together. Abi–Ezzi and Wozny have used the notation of *W–wraparound* in their recent paper ([Abie90]), a notation which can be considered as being more intuitive than the (although more widespread) notion of external lines. Both notations will be used in this thesis.

As the previous analysis on the appearance of external lines shows, the W–wraparound is a completely normal and well describable effect of projective geometry. It is therefore disappointing that this problem has not been widely addressed by the computer graphics textbooks. In fact, none of the traditional works on computer graphics ([Newm79], [Fole84], [Salm87] or even the brand new [Watt89]) mention the existence of the problem at all. It is therefore not surprising that a number of commercially available 3D graphics systems fail to handle the problem correctly; at the time of writing, some of these systems would just draw the line segment $P''Q''$ as the image of PQ. The description given above explains why

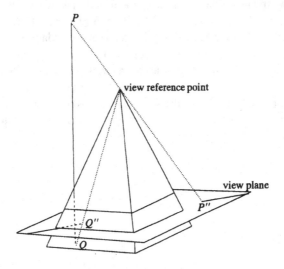

Figure 3.3.

this may happen without being noticed: as long as a very traditional synthetic camera model is used, and the object to be visualised tends to be more or less within the view volume (that is far away from the view reference point and hence from the vanishing plane) no problem will occur; the system will work perfectly.

Precisely handling the W–wraparound in a 3D system implementation is however a necessity. As mentioned before, if the system is used for the visualisation of objects which are very "close" to the observer, the wraparound might occur (see figure 3.3 again). The visual effect on the screen will be the appearance of, at least at a first glance, inexplicable line segments on the screen. Knowing the projective geometrical background, one may realise that these line segments are the "complementary" ones; the effect on the screen is, however, very disturbing. If, by chance, some of the objects to be visualised happen to have some points on the vanishing plane itself, the objects will seem to "blow–up" on the screen (or the program will even fail). This is also a method to check whether an actual implementation is handling the problem properly: one has to put simple objects into the 3D scene and make a program which "moves" the view reference point of the projection among the objects. In a number of commercial PHIGS implementations, for example, such a simple test program will produce irregular lines on the screen. This also means, that a proper algorithm to handle the W–wraparound is *necessary* for a proper implementation of the 3D output pipeline.

Just as the references to the W–wraparounds are missing from the usual textbooks, not too much has been published about the means to manage it properly. In fact, the paper of J.F. Blinn and M.E. Newell, published 1978 ([Blinn78]) was about the only known method for a fairly long time. In this paper, Blinn and Newell reformulate the clipping against the view volume for homogeneous coordinates, making use of some results published even earlier in [Suth74]. The problem is that the method presented in [Blinn78] had been (admittedly) developed with exclusively the synthetic camera model in mind and failed to be usable for a general case. However, for a long time, this model was practically the only one in use, and therefore this approach fulfilled the requirements and there seemed to be no need for any further development.

The appearance of the ISO 3D standards (or standard proposals), that is first GKS–3D ([ISO88]) and then PHIGS and its derivatives ([ISO89], [ISO88b], [ISO89a]) together with the first development projects aiming at the implementation of these specifications gave a new impulse to the solution of these problems. It was in 1987 that the so–called *W–clip* was first published; it was in use in two independent and parallel developments in Europe. One was the implementation of the full GKS–3D standard by the firm Insotec Consult GmbH in Munich under the name of GKSI (this is the implementation in which the author participated, see also [Herm87][t]) while the other one was the so–called KRT^3 project, which aimed at a PHIGS implementation at the University of Manchester ([Howa87], [Hubb87]). Both of them came to the same results, although in case of GKSI the far–reaching consequences of the basic approach in use were much more exploited than in case of the KRT^3 project (more examples of that will be presented later in this chapter). However, strictly for the problem of external lines, the two approaches were essentially the same. (As far as one can draw consequences out of the reference list of other publications, the idea has also been reused since then by other PHIGS and PHIGS PLUS implementations, see for example [OBar89] or [Abie90]). It was in 1989 that Krammer published (in [Kram89]) an alternative approach, called the UW–clip (see later); presumably this method had been used for a PHIGS implementation called IXPHIGS and realised at the Computer and Automation Institute of Budapest, described also in an earlier paper ([Görö88]).

3.1.2. The W–Clip

It is difficult to understand why the W–clip method was published in 1987 only; once the underlying projective geometry principles of the transformation process are really understood, the method seems to be just trivial. The only explanation seems to be the lack of projective geometry in the usual computer graphics curriculum, that is most of the implementors were not aware of the existence of the theory, let alone the details of it.

Figure 3.1 gives a clue of what can be done in 2D. The reason for the appearance of external lines is the fact that the image of the line segment *PQ* includes an ideal point as well, which disappears when displaying the affine points. However, this situation can be described in fully Euclidean terms as well: it is equivalent to the

---

[t]The name of this implementation has been recently changed to "DIGI–GKS" as a result of the fact that the firm DIGIDATA mbH has acquired the program, marketing it primarily in Germany.

fact that the line segment $P'Q'$ intersects (in Ψ) the $w=0$ plane. In fact, these are the points in $\mathbb{R}^3$ which represent (in homogeneous form) the ideal points of IPE^2. The possible solution comes as a result of this observation: before performing the projective division, a traditional clip has to be applied to the line segment $P'Q'$ to get rid of the potentially dangerous points. Clipping (that is cutting the invisible part of a geometric primitive or) is a very well described algorithm in computer graphics for simple clipping areas like half–planes or half–spaces; in other words, performing this clip means making use of a number of already existing algorithms (listed in numerous textbooks or tutorials like [Mudu86]). To be on the safe side, the effective clipping should be made for the half–planes $w \geq \varepsilon$ and $w \leq -\varepsilon$, where ε is a small positive real number.

The three dimensional case is similar. Ψ is now a three dimensional sub–space (that is a hyperspace) of $\mathbb{R}^4$; otherwise the analogy to the 2D case is complete. Of course, one should check whether the clipping algorithms in use to perform the $w \geq \varepsilon$ as well as the $w \leq -\varepsilon$ clip are applicable in 4D as well; however, all such algorithms use as an elementary calculation step the fact that a line segment PQ can be described by the formula

$$tP + (1 - t)Q \qquad 0 \leq t \leq 1 \tag{3.1}$$

and that a plane in 3D (or a line in 2D) can be described in Cartesian coordinates by the equation

$$\{x : x^T n + \alpha = 0\} \tag{3.2}$$

where n is the normal vector of the plane/line and α is essentially the distance of the plane/line measured from the origin. Using these formulae, the (eventual) intersection point of the line segment and a plane can be calculated easily and hence the clipping problem becomes a programming problem rather than an algorithmic one (the program should keep lists of points which are "outside" or "inside" etc).

Formulae in 4D are not really different. Formula (3.1) still describes a line segment in $\mathbb{R}^4$ and a hyperspace can be described by formula (3.2), with the parameters having the same meaning as in 3D or 2D. Consequently, all the traditional clipping algorithms can be adapted without additional difficulties for the four dimensional space as well.

The above process has been given the name *W–clip*. The W–clip is definable for all graphics primitives and not for line segments only, in spite of the fact that these have been used to clarify the way the whole approach can be introduced. The same clipping can be done (and, in fact, should be done) for polygons or for more complicated geometric primitives, although most of the algorithms use, in practice, line clipping algorithms internally (for example to clip on the edges to determine the clipped part of the polygon).

It should be clear by now that the real clue to the usability of the W–clip is the fact that a projective transformation can be viewed as a two–stage process: first the linear part has to be performed and then the projective division. Nothing prevents the implementor of the transformation from inserting something between these two steps and this is exactly what is done in the W–clip. Also, by using this intermediate stage, a number of calculations can be performed in the usual Euclidean environment; the only additional price to be paid is that four dimensional rather than three dimensional geometry should be used[†].

The W–clip, as described above, is a two–stage clip: a separate clip has to be performed against the $w \geq \varepsilon$ and the $w \leq -\varepsilon$ half–spaces. As clipping is a relatively time–consuming operation, it is of course an important question to see whether one of the two clips can be avoided to reduce the complexity of the W–clip.

First of all, no W–clip is necessary in the case of an affine viewing (that is a parallel projection). This fact is trivially true from the definition of an affine transformation; furthermore, by inspecting the last row of the transformation matrix it can be determined whether the viewing in use is affine or not.

If the implementation is designed for the GKS–3D pipeline then, conceptually, there is a clip called the *normalisation clip* before viewing. This clip is essentially a traditional clip against a cube which is enclosed in the $[0,1]^3$ cube of $I\!R^3$. The implementor might choose to perform this clip prior to viewing (although, by taking the normalisation clip to be a special case of the modelling clip, this step can be postponed after the viewing as well, see chapter 4). The net result is that all points have as Cartesian (and hence homogeneous) coordinates non–negative values. It is trivial, therefore, that the following statement is valid:

Theorem 3.1. If $T = (T_{i,j})_{i,j=1}^{4}$ is the view matrix and all points to be transformed have non–negative coordinates only (eg in case of a GKS–3D implementation performing the normalisation clip prior to viewing) and the following relations are true:

$$T_{4,4} > 0 \text{ and } T_{4,j} \geq 0 \ (1 \leq j \leq 3) \quad \text{or}$$
$$T_{4,4} < 0 \text{ and } T_{4,j} \leq 0 \ (1 \leq j \leq 3) \tag{3.3}$$

then only one half of the W–clip is necessary ($w \geq \varepsilon$ for the first case and $w \leq -\varepsilon$ for the second one)[‡].

Clearly, the conditions ensure that all the points in Ψ which are of interest will have a $w > 0$ (resp. $w < 0$) coordinate value in $I\!R^4$. ∎

A much more important and more interesting optimisation case is as follows. In both basic 3D standards as well as in most 3D systems in general, viewing itself is followed by yet another clipping step, usually termed "workstation clip". Its

---

[†] Even this fact can be eased: what happens is that most of the algorithms which should be performed are defined for primitives within Ψ, that is a three dimensional subspace of $I\!R^4$, where the three dimensional geometry is locally still valid.

[‡] Care should be taken, however, when using this optimisation: in GKS–3D, the user has the option of switching the normalisation clip off, which means that the first premise of 3.1 is not valid any more.

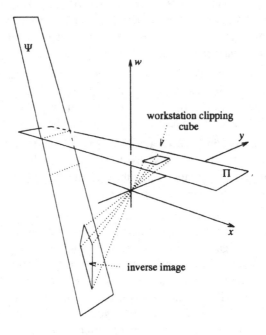

Figure 3.4.

purpose is to confine the output to one specific area of the visible screen by specifying a "regular" cube (that is a cube with sides parallel to the main axes). This cube is usually defined by giving its two diagonally opposite vertices (and, in fact, is the intersection of the so–called workstation window and the view viewport). This cube will be called the *workstation clipping cube* in the following discussion. The idea is to project the eight vertices of this cube back onto Ψ by the *inverse mapping* of the projective division (see figure 3.4; of course, being the analogous case for 2D, there are only four vertices in the figure instead of the eight ones for 3D) and to see whether all inverse images are on the same side of the $w=0$ hyperspace or not. Clearly, if they are, then the convex body determined in Ψ (which is a three dimensional subspace of $\mathbb{R}^4$) will lie completely on the same side of $w=0$ and, furthermore, this convex body will be the inverse image of the workstation clipping cube. This means that the *other* side of $w=0$ can be disregarded; any primitive clipped to it would be cut in a later step by the workstation clip anyway. One must be cautious, however: in some cases it might happen that the inverse image of one of the vertices is not on Ψ (the projecting line is parallel to Ψ or, in other words, the inverse image is an ideal point of the hyperspace Ψ). If this is the case, this particular optimisation step cannot be used.

The question now becomes how to describe the inverse of the projective division. Here again, the two dimensional analogy might help. First, a vector which is

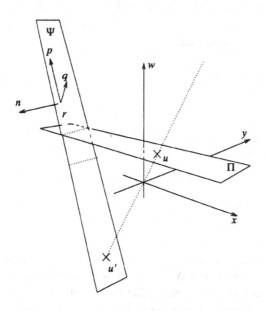

Figure 3.5.

perpendicular to Ψ is necessary. In the 3D case, a possibility is to take three non-collinear (Euclidean) points on Ψ (these can be found by transforming three non-collinear points of Π by the projective transformation in use); these points are denoted by p, q and r. Clearly

$$n = (\vec{p} - \vec{r}) \times (\vec{q} - \vec{r}) \tag{3.4}$$

will be perpendicular to Ψ.

If the point $u = (u_1, u_2, 1)^T$ is on the plane Π, the points on the projecting ray (in fact, the homogeneous coordinates) can be described by

$$tu = (tu_1, tu_2, t) \qquad t \in \mathbb{R} \tag{3.5}$$

Finding the intersection u' of this ray with Ψ means finding a $t \in \mathbb{R}$ where:

$$tu^T n = r^T n \tag{3.6}$$

which gives the inverse of the projective division.

The only problem when generalising this procedure for 4D is to find an analogous formula for (3.4). The main point in using (3.4) is that the outer product has the property of being perpendicular to both of its multiplicands. This is what has to be generalised for higher dimensions.

In the case of 4D not three but four points on Ψ will be necessary which are of a general position, that is no three of them are collinear. If these points are denoted by p, q, r and s, it is necessary to find a vector which is perpendicular to all three vectors $(\vec{p}-\vec{r})$, $(\vec{q}-\vec{r})$ and $(\vec{s}-\vec{r})$. In the 3D case, the outer product of two vectors $a,b \in I\!R^3$ can be calculated by:

$$a \times b = \det \begin{bmatrix} a_1 & a_2 & a_3 \\ b_1 & b_2 & b_3 \\ e_1 & e_2 & e_3 \end{bmatrix} \tag{3.7}$$

It is a trivial algebraic calculation to show that if $a,b,c \in I\!R^4$ and the outer (or vector) product of these vectors is defined by:

$$a \times b \times c = \det \begin{bmatrix} a_1 & a_2 & a_3 & a_4 \\ b_1 & b_2 & b_3 & b_4 \\ c_1 & c_2 & c_3 & c_4 \\ e_1 & e_2 & e_3 & e_4 \end{bmatrix} \tag{3.8}$$

then the resulting vector will be perpendicular to a, b and c. Therefore, by the substitution of formula (3.8) into (3.4), the inverse of the projective division can be calculated easily. It has also to be stressed, that this calculation only requires to be performed once for a given transformation and workstation clipping cube, and it is not dependent on the output primitive being treated.

It is interesting and deserves a detour to examine how useful this optimisation really is. In other words, it is valid to ask how frequently will formula (3.6) lead to positive and practical results for the workstation clipping cube.

The value of $r^T n$ can be considered as being positive; if this were not the case, $-n$ could be taken instead (the case when $r^T n = 0$ has been deliberately disregarded as a very special case). This means, that the test which has to be performed can also be formulated as follows: is it true that for all vertices u_i of the workstation clipping cube the values of

$$u_i^T n \tag{3.9}$$

are of equal sign? In fact, in (3.6) only the sign and not the exact value of t is really of interest in deciding whether to use both halves of the W–clip or not.

In the usual Euclidean environment (both in $I\!R^3$ and $I\!R^4$) formula (3.9) is equivalent to the question whether all u_i vertices are in the same half–space determined by the hyperspace Ψ', where Ψ' is the plane/hyperspace in $I\!R^3/I\!R^4$ which crosses the origin and whose normal vector is parallel to n (see figure 3.6). As all vertices u_i are points of Π as well, this can be reformulated by saying that (3.9) is equivalent to the question whether all u_i vertices are in the same half–space of Π determined by $\Pi \wedge \Psi'$. In other words, the following statement has been proven:

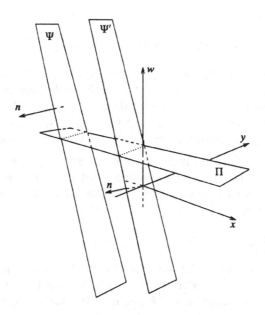

Figure 3.6.

Theorem 3.2. The W–clip optimisation based on the inverse projection of the workstation clipping cube leads to reducing the number of clipping steps if and only if the cube is not intersected by the plane $\Pi \wedge \Psi'^{\dagger}$.

But what is exactly $\Pi \wedge \Psi'$? Looking at figure 3.6 it is clear that if T is the transformation in use then:

$$\Pi \wedge \Psi' = T(\mathit{I})\tag{3.10}$$

that is, this plane is the *image of the ideal plane*. Thus, theorem 3.2 says that the optimisation step will lead to a positive result if and only if the image of the ideal plane does not intersect the workstation clipping cube.

Let have a look at the traditional synthetic camera model. If the front and back clipping planes of the view frustrum are denoted by Ω_1 and Ω_2, and the vanishing plane of the transformation T is Θ (remember that this plane is parallel to Ω_1 and Ω_2 and crosses the view reference point), then the planes Ω_1, Ω_2, Θ and I form a bunch of planes which have as a common intersection line an ideal line (the first three planes are indeed parallel). Furthermore, the domain cutting property applied to these planes says that:

[†]This also means that an alternative way of performing the optimisation would be to check this fact directly; however, the necessary formulae involved would be of the same complexity.

$$(\Omega_1 \Omega_2 \Theta \mathit{\Pi}) > 0 \qquad (3.11)$$

Consequently:

$$(T(\Omega_1)T(\Omega_2)T(\Theta)T(\mathit{\Pi})) > 0 \qquad (3.12)$$

$T(\Omega_1)$ and $T(\Omega_2)$ will be the two parallel planes with $z=max$ and $z=min$ for the workstation clipping cube, $T(\Theta)$ will be (by definition) the ideal plane and, finally, $T(\mathit{\Pi})$ is the plane which is used in the formulation of theorem 3.2. Furthermore, the domain cutting property (and formula (3.12)) says that $T(\mathit{\Pi})$ does *not* intersect the workstation clipping cube. In other words the following statement has been proven:

> **Theorem 3.3.** If T is the transformation corresponding to the synthetic camera model, the W–clip optimisation based on the inverse image of the workstation clipping cube will enable the reduction of the number of clipping steps.

3.1.3. The UW–Clip

The UW–clip, introduced by Krammer in [Kram89] uses a completely different approach for handling the W–wraparound problem. His aim is to find an appropriate clipping area in $I\!R^3$ which can be used *before* the effective viewing to remove potentially dangerous points from the output primitives. This approach will be presented here with slightly modified arguments for its description.

The basic idea is very close to the one used in proving theorem 3.3. The vanishing plane of the transformation T (denoted again by Θ) generates a bunch of planes in $I\!P\!E^3$: the set of all planes parallel to Θ plus $\mathit{\Pi}$ itself. All these planes have a common intersecting line (which, belonging also to $\mathit{\Pi}$, is an ideal line). The transformation T maps this bunch of planes onto a bunch of planes again; this latter consists of parallel planes (in a Euclidean sense). Indeed, as the image of Θ is the ideal plane, the intersection of the images is an ideal line, which is equivalent to the fact that the planes are parallel.

Figure 3.7 shows the analogous situation in 2D using lines instead of planes. It is also straightforward to find the (Euclidean) equation for all planes in the bunch of image planes: the images of the points $[(1,0,0)]^T$, $[(0,1,0)]^T$ and $[(0,0,1)]^T$ can be used to describe the homogeneous coordinates of $\mathit{\Pi}'=T(\mathit{\Pi})$ (see formula (2.21)), and these homogeneous coordinates result in the series of equations of the form

$$ax + by + cz + \alpha = 0 \qquad (3.13)$$

where α serves to differentiate among the different elements of the bunch of planes.

The planes U' and V' are chosen from this bunch so that the stripe $U'V'$ contains the workstation clipping cube (see again figure 3.7). These planes, being elements of the bunch described by (3.13), can be transformed by T^{-1} onto two planes U and V which are parallel to Θ. The planes U and V determine a conic sector in

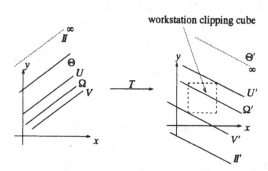

Figure 3.7.

$I\!\!P\!E^3$. As far as the affine points are concerned, one of the two sectors is the stripe defined by U and V and the other is its complement.

Conic sectors are mapped onto conic sectors; therefore either the UV stripe or its complement is mapped by T onto the stripe $U'V'$. Using essentially the same arguments as for theorem 3.3 the following can be proven:

Theorem 3.4. If $I\!\!I'$ does not intersect the workstation clipping cube, the image of the stripe UV will be $U'V'$, its complement otherwise. Additionally, if the first situation is encountered (that is $I\!\!I'$ does not intersect the workstation clipping cube), it is also true that Θ does not intersect the stripe UV. On the other hand, if $I\!\!I'$ does intersect the $U'V'$ stripe, Θ will be part of the stripe UV.

Suppose that $I\!\!I'$ does not intersect the workstation clipping cube (as in figure 3.7). If Ω' is an arbitrary line which is parallel with U' and V' and, furthermore, it intersects the workstation clipping cube, the following holds:

$$(I\!\!I'\Omega'U'V') < 0 \qquad\qquad (3.14)$$

Consequently,

$$(I\!\!I\Omega UV) < 0 \qquad\qquad (3.15)$$

which is possible if and only if Ω runs within the strip UV. As far as the second statement is concerned, if $I\!\!I'$ is disjoint from the stripe $U'V'$, then

$$(\Theta'I\!\!I'U'V') > 0 \qquad\qquad (3.16)$$

(Θ' is the ideal plane), that is

$$(\Theta I\!\!I UV) > 0 \qquad\qquad (3.17)$$

which is possible if and only if the plane Θ does not intersect the strip UV. ∎

The UW–clip consists therefore in determining the planes U and V and in performing a clip in $I\!R^3$, prior to the transformation itself, against either the stripe UV or against the complement of it. Which of the two possibilities is to be chosen is decided by theorem 3.4. This algorithm is indeed correct; if, say, the image of the UV stripe is the $U'V'$ stripe, then all affine points in this stripe will be mapped onto affine points again (Θ being the collection of the points which are mapped onto ideal ones!).

3.1.4. Comparison of the W–Clip and the UW–Clip

Mathematically speaking (that is from a mathematician's point of view) the UW–clip is much more elegant. Although no mathematician can ever define properly what the "elegance" or the "beauty" of a mathematical theorem, proof or construction really means, all of them have an indescribable feeling for it; the fact that the UW–clip makes use of a clean projective geometrical construction only instead of a mixture of projective and Euclidean geometry (as the case for the W–clip) makes it more conform to a traditional mathematical approach. However, programmers are more concerned about efficiency and other practical problems and therefore such an argument is less significant in this case.

It is difficult to give an exact algorithmic comparison of the two methods. They seem to be of similar algorithmic complexity and it is therefore the actual computing environment which might influence the final choice. As stated previously, both algorithms have been implemented in the course of independent development projects, resulting in competing products. There has been no opportunity to perform any tests to compare them in an objective way.

The UW–clip has the undeniable advantage of performing a clip prior to the transformation itself and, in consequence, eventually reducing the amount of matrix–vector multiplication to be done. For a fully software–based implementation this fact may have great importance, the matrix–vector multiplication being a computationally demanding step. However, the clips themselves are generally more complex than in case of the W–clip: while in the case of a UW–clip two planes in $I\!R^3$ are to be used, the position of which may be arbitrary in space, for the W–clip the planes involved ($w=\varepsilon$ and $w=-\varepsilon$) are very simple to handle. This also means that the latter is much easier to implement both in software and eventually in special hardware. As an example, a simplified version of the W–clip has been implemented by the author on special graphics hardware based on dataflow techniques ([Hage90]); as the hardware used had a very simple instruction set, it would not have been possible to implement a full UW–clip on it, due to the lack of necessary instructions[†]. Also, the real value of avoiding a larger number of matrix–vector

---

[†]The graphics machine described in [Hage90] uses 3D triangles as basic primitives to approximate 3D surfaces. Due to the lack of a division instruction, the triangles could not be clipped; instead, all those triangles which had a change in sign for the w values of their vertices were just disregarded in the rendering process.

multiplications depends on the actual environment: it is quite frequent nowadays to have special hardware or firmware to perform this operation (see again [Hage90]) which makes it therefore relatively "cheap", so that its simplicity might become a decisive argument in favour of the W–clip.

The UW–clip depends quite heavily on the existence of a workstation clipping cube whereas in case of the W–clip this is just the source of a possible optimisation. Credit should be given to the fact that practically all 3D systems define such a cube somewhere in the pipeline, which makes the UW–clip also fairly general.

Probably the greatest advantage of the W–clip is that it has created interest in the possible advantages of using the four dimensional space for the purposes of computer graphics and this has resulted in a series of new approaches. A number of these (namely the implementation of cell array, new pattern filling and stroke text generation algorithms) have also become an integral part of the commercial GKS–3D implementation GKSI/DIGI–GKS mentioned before, while others have been tried out in experimental cases only; however, all of them make use of the fact that the clipping step (which may destroy a number of regular features of the output primitives) is performed relatively late in the pipeline. This is what will be presented in what follows.

3.2. Linear Primitives in $\mathbb{R}^4$

3.2.1. Cell Array

One of the output primitives which is a source of many implementation problems for graphics standards is the *cell array*. The reason for the difficulties is that in its case speed is of an overall importance; if not implemented efficiently the primitive becomes virtually unusable. While this statement is of course true for all primitives, for cell arrays one has to deal with large (in the range of hundreds times hundreds) arrays of coloured cells; in other words, even a slight change for the worse or for the better in the implementation algorithms might sum to a very significant change when using the primitive in practice.

The idea of a cell array is very simple. It is essentially the general form of a pixel image, that is of an image consisting of a series of individual cells, each of them being assigned an appropriate colour. It is one of the rare output primitives which have been added to all the ISO graphics standards to include raster–like pictures. It is, at least theoretically, a potentially very powerful primitive: for example, it allows the user of the graphics package to include digitised pictures into the output stream together with the geometric primitives. However, it is exactly this feature which makes it difficult to implement: the number of pixels might be huge, so that even if the primitive can be rendered by a series of simple individual steps, the large number of such steps may required significant amounts of processing time.

A cell array is defined in the coordinate space in use prior to any transformation (that is in world coordinate space in GKS, modelling coordinate space in PHIGS) by determining the corners of a rectangular array, traditionally denoted by

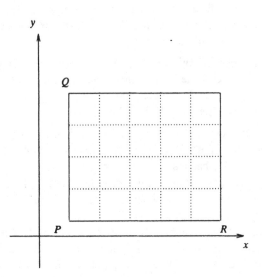

Figure 3.8.

P, Q and $R^\dagger$. The edges of the parallelogram are then divided by equal steps into a number of sub–parallelograms where the number of divisions has to be specified by the user. The result is a regular grid, as shown in figure 3.8. Finally, each sub–parallelogram is assigned a colour. To ensure compatibility with the remaining output primitives, a cell array is subject to all the transformations defined in the output pipeline; that is the grid shown in figure 3.8 can be distorted before being displayed on the screen.

As already mentioned, the difficulty is that the number of subdivisions can be very high which results in a very large number of internal sub–parallelograms. Conceptually, each of these sub–parallelograms has to be transformed individually and displayed on the screen as some kind of small polygon to be filled by a given colour.

In a 3D system, the points P, Q and R are, of course, points in $\mathbb{R}^3$. The vectors $\vec{R} - \vec{P}$ and $\vec{Q} - \vec{P}$ determine a plane in $\mathbb{R}^3$; the cell array is (conceptually) a pixel image on this plane. In the course of the output pipeline, this parallelogram in space is transformed by the viewing transformation in use to result in distorted pictures, as for example the one shown in figure 3.9.

The distortions shown in figure 3.9 are the real source of the algorithmic difficulties. In a 2D system, the maximum possible distortion produced by the transformations is the creation of parallelograms instead of rectangles: the

---

†In the case of GKS/PHIGS, only a rectangular array is allowed, whereas in CGI ([ISO88a]) a general parallelogram can also be defined.

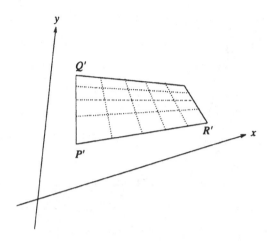

Figure 3.9.

transformations used in such a case are always affine, that is parallelograms will be transformed into parallelograms (the edges, which have an ideal intersection point on the original image, will still have ideal intersections after the transformation, that is the image must be a parallelogram). Furthermore, the equal subdivision of the edges will still remain an equal subdivision afterwards (affine transformations maintain the division ratio!). Consequently, an implementation may refrain from transforming all internal sub–parallelograms: it is perfectly feasible to transform the points P, Q and R only and apply the division procedure only afterwards.

All of these arguments are far from true in the case of a non–affine transformation. The image of the parallelogram may become a general quadrilateral and the internal subdivision will also be distorted. In other words, it is *not* possible to transform just the points P, Q and R; these data do not contain enough information any more to reconstruct the image of the original parallelogram after the transformation.

What can be done? The usual approach is to say that the implementation is simply forced to construct the appropriate parallelogram in space prior to the transformation: this is the only way to achieve the projective distortions like that in figure 3.9. This means, in practice, that this primitive becomes too slow to use. In spite of this, almost all implementations follow this approach. Fortunately, there are ways to overcome this problem, even several. One of them will presented in what follows (and has originally been presented in [Herm87]), while another will be presented in chapter 5 (based on results originally published in [Herm89]).

The first solution is based on the same concepts as the W–clip and is illustrated (again in 2D) in figure 3.10. The idea is as follows. If the transformation in use is denoted by T, its linear part (that is the first step in performing the

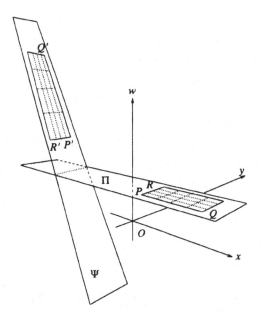

Figure 3.10.

transformation) is a linear transformation in $\mathbb{R}^4$ ($\mathbb{R}^3$ on figure 3.10). It can also be considered as being an affine transformation which transforms the hyperspace Π onto Ψ. This affine transformation maps the points P, Q and R onto the points P', Q' and R'. Furthermore, being affine, it will also automatically map the parallelogram determined by P, Q and R onto the parallelogram defined by P', Q' and R' (see the previous considerations). In other words, only the projective division is responsible for the distortion of the final image.

The way of handling a cell array is therefore based on the idea of performing the subdivision of the hyperspace Ψ *after* having performed the matrix–vector multiplications on the points P, Q and R but *before* the projective division. Based on the nature of projective transformations, the result will be the same.

Performing the calculations in 4D does not create any difficulties. To get the internal subdivision points the (4 dimensional) vector equations must be used:

$$\vec{P} + i\frac{dist\,(PR)}{n}(\vec{R} - \vec{P}) \quad (1 \le i \le n-1) \tag{3.18}$$

for the internal subdivision point on the edge PR and

$$\vec{P} + j\frac{dist\,(PQ)}{m}(\vec{Q} - \vec{P}) \quad (1 \le j \le m-1) \tag{3.19}$$

on the other side. n and m are the number of internal subdivisions. The only "four dimensional" feature in these formulae is how to calculate the distance of two points, however

$$dist(PR) = \sqrt{\sum_{i=1}^{4}(P_i - R_i)^2} \tag{3.20}$$

is valid in 4D as well.

The resulting sub–parallelograms are then handled separately, at least logically speaking (in programming practice, one can reduce the number of divisions by making use of the shared points). Before performing the projective division, that is to get back onto Π, the W–clip has to be performed for each individual quadrilateral to avoid the appearance of ideal points in the results. Some additional techniques are available to ease this step as well: if, for example, all vertices of the full parallelogram are on one side of the $w=0$ plane, then no W–clip is necessary at all. Indeed the parallelogram is a convex set of $I\!R^4$; that is if all four vertices are in either $w>0$ or in $w<0$, the image of the whole parallelogram is also automatically disjoint from ideal points. If this pre–test fails, the test can still be done for each individual sub–parallelogram again.

Clearly, this approach is significantly faster than the "straightforward" method: no matrix–vector multiplications have to be done on the internal points. Its importance lies also in the fact that it proves the usability of a general approach: by making use of the two–stage characteristics of a projective transformation, a number of algorithms can be performed in 4D rather than in 3D without altering the result but with a significant gain in speed. Some of the problems are very similar to the cell array whilst some of them need additional considerations.

3.2.2. Pattern Filling

In case of pattern filling of a polygon, a rectangular pattern is defined in very much the same way as a cell array with the further complication that the pattern is defined *on the plane of the polygon*. This generated pattern is then extruded through the polygon itself. The result is what is called a polygon filled with a pattern interior style (for further details of the specification, the reader should refer to the relevant and already cited ISO documents). Pattern filling is similar to cell array, as far as the difficulties related to projective transformations are concerned.

The problem is similar: the pattern can be distorted by the projective transformation, and therefore, to achieve a true three dimensional effect, the pattern filling itself cannot be performed simply after the full projective transformation. However, the pattern can be reconstructed, just like the case of a cell array, in four dimensional space, using formulae (3.18) and (3.19) respectively. The difference is that the indices i and j in these formulae may run (conceptually) from $-\infty$ to ∞. Hence, it is now a natural idea to perform the pattern filling process using 4D data instead of 3D to try to improve the traditional approach.

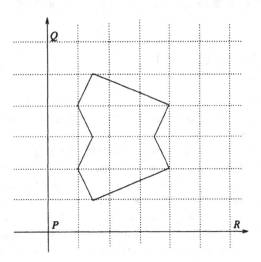

Figure 3.11.

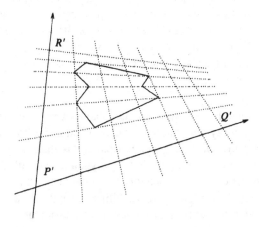

Figure 3.12.

Care should be taken, however, when generalising the procedure of pattern filling; the question is whether the pattern filling itself can be done or not in 4D. In other words, do the traditional approaches to perform pattern filling apply in the four dimensional case? If yes, is it worthwhile to do it?

One of the usual algorithms to perform pattern filling is by using yet another (temporary) transformation, which is a coordinate transformation (of Cartesian coordinates). A local coordinate system is chosen, which has as its main axes the two main edges of the pattern parallelogram ($P'Q'$ and $P'R'$ on figure 3.12) and which would therefore transform the problem back into a regular situation shown in 3.11.

How this transformation has to be chosen is a standard procedure in linear algebra. If $a = \vec{Q} - \vec{P}$ and $b = \vec{R} - \vec{P}$, then let the matrix M be as follows:

$$M = \begin{pmatrix} a_1 & b_1 & 0 & 0 \\ a_2 & b_2 & 0 & 0 \\ a_3 & b_3 & 1 & 0 \\ a_4 & b_4 & 0 & 1 \end{pmatrix} \tag{3.21}$$

Clearly, $Me_1=a$, $Me_2=b$, $Me_3=e_3$ and, finally, $Me_4=e_4$. Consequently, M^{-1} will be the necessary coordinate transformation. Once this transformation has been performed, the pattern filling itself (ie clipping the sub–parallelograms against the polygon etc) becomes a fairly "classical" procedure to be performed, based on well–documented algorithms. The resulting clipped quadrilaterals are transformed again by M.

The necessity of using this additional transformation is, however, not very attractive; it is therefore worth examining whether it eventually spoils the advantages of the whole 4D approach.

To answer this question the necessary steps are listed for pattern filling. For the 4D case they are as follows.

i) The polygon is transformed, using matrix M^{-1} (M is the matrix of formula (3.21)). The result is a local two dimensional environment, like that in figure 3.11.

ii) The sub–parallelograms of the defined patterns are clipped against the (transformed) polygon to generate a series of small polygons.

iii) Each polygon is transformed by M and rendered on the output medium.

When the procedure is performed in 3D, the points generated in step iii) must still be transformed by the full viewing transformation, that is instead of the three dimensional counterpart of M the transformation TM should be used, where T is the viewing transformation. For a 4D version, T has already been (conceptually) applied, so M alone is needed. Consequently, the algorithmic difference between the 3D and the 4D version can be described by the fact that the matrix to be used is different: M is a relatively sparse matrix in 4D and, furthermore, the third and the fourth columns can be disregarded, whereas no assumption can be made on the form of TM. In the course of the development of the already mentioned GKSI system, both versions have been implemented to compare the results. The improvement of the 4D version was very clear: a speed increase between 25% to 30% (depending on the pattern size, of course) has been achieved on an Apollo DN3000 Workstation (the measurement data were related to the whole output, including the drawing itself; comparing the quadrilateral calculations only, the result might have been even better).

It has to be stressed that the above described method is not the only one available to perform pattern filling. By doing calculations in Ψ, the limits in the indices i and j from formulae (3.18) and (3.19) can be found directly as well. What has to be examined is whether the whole polygon is in one half–space of the (three dimensional) hyperspace Ψ by making also use of the fact that the polygon and the corresponding pattern are in the same plane (by definition). These calculations can be made: the analogous three dimensional calculations can be generalised easily for 4D. Once a finite part of the whole pattern array has been found, pattern filling is reduced to traditional clipping again: as mentioned before, this problem (which is usually based on performing a line segment intersection calculation) can be done in 4D as well. The resulting formulae are, however, quite complicated and computationally demanding; in fact, the result is comparable to the approach based on coordinate transformation.

3.2.3. STROKE Characters

The last problem falling into the same category is the generation of STROKE (or high precision) characters.

When implementing STROKE characters in a general graphics system, the characters themselves are usually described internally on some kind of grid, or integer valued Cartesian coordinate system. The exact resolution of this grid is dependent on the environment and is usually hidden from the user of the graphics system. However, this number exists (it has typically a value around 100). Let us consider for the time being that this value is k (both vertically and horizontally).

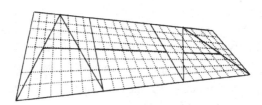

Figure 3.13.

Furthermore, text drawn in stroke precision is defined as a succession of planar character boxes (all boxes lying on the same plane in space). For the sake of simplicity, only the situation where these boxes are of the same size is considered. These two regular grids (the succession of character boxes and the character description grid) form a larger grid on the whole text extent parallelogram, which has a resolution of k in the "vertical" direction (as a result of the character description) and a resolution km in the "horizontal" direction, where m is the number of characters in the text. The grid generated for the string "AHA" is presented in figure 3.13; to avoid confusion, however, a much coarser grid is shown, with a value of $k=8$ instead of the order of 100.

The problem is therefore very much the same as in the case of cell array and pattern filling: the regular grid is distorted by the transformation to achieve the visual appearance of figure 3.13. The problem being the same, the same solution can also be applied; in contrast to the case of pattern filling no additional difficulties for the generation of characters arise. Using formulae (3.18) and (3.19) the whole grid of figure 3.13 can be reconstructed on the hyperspace Ψ and this grid is subsequently used to generate the characters themselves.

3.3. Conics

3.3.1. Introduction

Some of the problems concerning conics have already been presented in an earlier chapter. Essentially, each conic has to be replaced somewhere along the output pipeline by an appropriate polygon or polyline to approximate the complete curve or an arc of it. The number of points in the approximating polygon or polyline is relatively large; consequently and for obvious reasons, it is very important to postpone this replacement to reduce processing time.

Conics are not linear primitives like cell arrays, patterns or STROKE characters; in other words the ideas presented concerning these latter primitives cannot be mechanically reused. However the basic approach, that is to perform some of the necessary algorithms after the linear part of the view transformation but before the projective division does work for conics as well; this will be presented in the present chapter.

What is necessary is to have an *affine invariant* description for each class of conics; the linear part of the transformation being an affine transformation of two hyperplanes of $I\!R^4$, such a description may provide the necessary tool needed to postpone the generation of the approximating polygon/polyline into 4D. In general, the approach will be as follows.

Suppose for each class of conics (that is for ellipses, hyperbolae and parabolae) there exists a set of points $C_1, C_2, \ldots, C_n$, (called the set of *characteristic points* of the conic) and a general function ϕ so that the following two statements are true:

- The conic (denoted by C) can be described with the help of the function ϕ:

$$C = \{ \phi(C_1, C_2, \ldots, C_n, t) : t \in I \subset I\!R \} \tag{3.22}$$

where I is a finite parameter interval of $I\!R$ (usually $[0, 2\pi]$). In other words, if the characteristic points are fixed, ϕ is a function which maps the interval I into $I\!R^n$ and the image of I is the conic itself.

• The function ϕ is affine invariant. This means that if T is an affine transformation, then the following is true:

$$T(C) = \{ \phi(T(C_1), T(C_2), \ldots, T(C_n), t) : t \in I \subset I\!R \} \qquad (3.23)$$

This means that the characteristic points effectively characterise the curve (at least in the affine sense): by transforming the characteristic points only, the transformed curve becomes fully describable. It has to be stressed that the choice of the characteristic points as well as the exact formulae describing ϕ depend on the class of the curve: the way to construct ϕ will be similar for all three cases but not the resulting formulae.

If such formulae and possible choices for characteristic points exist, then the generation of the curve can be postponed to 4D. For each curve the characteristic points $C_1, C_2, \ldots, C_n$ are transformed by the linear part of the transformation resulting in the set of characteristic points $C'_1, C'_2, \ldots, C'_n$ in the hyperspace Ψ; in fact, these points generate a two dimensional subplane of $I\!R^4$ (and hence of Ψ) which is the plane of the conic. The approximation of the curve can be done by choosing a discrete (but arbitrarily dense) subset of the parameter interval I; denoting these parameter values by $t_1, t_2, \ldots, t_n$, the points $\phi(C'_1, C'_2, \ldots, C'_n, t_i)$ will generate the necessary linear approximation of the curve. It is a very essential factor of the whole procedure that ϕ should be independent of the actual dimension (in fact, ϕ will always be a vector equation, involving vectors like $\overrightarrow{C_2} - \overrightarrow{C_1}$ and $\overrightarrow{C_3} - \overrightarrow{C_1}$). This method of curve generation is very powerful: just as in the case of cell array, the number of points for which the matrix–vector multiplication is to be performed can be significantly reduced, without losing the quality of the approximation.

A third, but also very important constraint for ϕ should be as follows:

• ϕ should be easily invertible. This means that if a point P on the curve is given, it should be possible to generate the value of $\tau \in I$ so that:

$$P = \phi(C_1, C_2, \ldots, C_n, \tau) \qquad (3.24)$$

This feature is very important to describe *arcs*; indeed, by giving three points of a conic arc, applying (3.24) means that the arc can be described as an appropriate subinterval of I.

The different functions ϕ for the different classes of conics will now be presented. The formula describing an ellipse is not new; in fact, it has been used previously in the CGI functional description ([ISO88a]). The corresponding formulae for parabolae and hyperbolae were however nowhere mentioned in the literature in their full generality; these had to be constructed based on some special (and already known) forms of equations to describe these curves. This has originally been published in [Herm89].

Two other general remarks are of interest. As shown in the next sections, all functions ϕ are of the form:

$$\phi(t) = \phi_1(t)(\vec{C_k} - \vec{C_1}) + \phi_2(t)(\vec{C_l} - \vec{C_1}) + \vec{C_1} \qquad (3.25)$$

where ϕ_1 and ϕ_2 are rational trigonometric functions, depending exclusively on the parameter t. These rational functions are not very friendly to calculate, so one possible objection to the general approach described above is that the complexity of these functions may jeopardise the advantages gained by the four dimensional approach. However, these functions do *not* depend on the actual curve; in other words, the values of these function for a subdivision of the parameter interval can be calculated in advance, stored in a look–up table and reused at run–time without loss of efficiency.

Another general remark is that for all classes of curves, as a "by–product", the general form of their matrix will also be generated. This fact does not necessarily have an importance as far as the four dimensional approach is concerned, but will be important later.

3.3.2. Affine Invariant Formulae

3.3.2.1. Ellipses

The simplest ellipse is a unit circle. There is also a well known parametric equation to describe the points of the circle, namely:

$$(cos\,(t), sin\,(t))^T \quad (0 < t \le 2\pi) \qquad (3.26)$$

The matrix describing the circle (as a conic) is also very simple, namely

$$A = \begin{bmatrix} 1 & 0 & 0 \\ 0 & 1 & 0 \\ 0 & 0 & -1 \end{bmatrix} \qquad (3.27)$$

The geometric features of the circle should be described in an affine invariant manner. Looking at figure 3.15 and comparing it with 3.14, one can deduce that the pole of the X axis is in the ideal point of the Y axis, that is $[(0,1,0)]^T$. Conversely, the pole of the Y axis is $(1,0,0)^T$, that is the ideal point of the X axis. (See also the description of the pole–polar relationships in the case of an external point of the curve in the previous chapter). In other words, the two coordinate axes form a conjugate pair of lines, more exactly a conjugate diameter pair (in other words, the radii CQ and CR form a pair of conjugate radii).

An affine transformation keeps the conjugate diameters (it keeps conjugation because it is a projective transformation and it keeps the centre because it is affine). Let T be the transformation which transforms the circle on figure 3.14 into the ellipse of figure 3.15 by $C \rightarrow C'$, $Q \rightarrow Q'$ and $R \rightarrow R'$. The result will be an ellipse (the ideal line does not change) and the lines $C'Q'$ and $C'R'$ will be a pair of conjugate

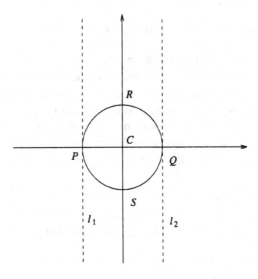

Figure 3.14.

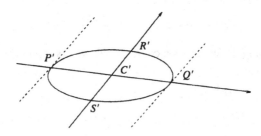

Figure 3.15.

radii. The matrix of the transformation is also straightforward: by denoting $u' = \vec{Q'} - \vec{C'}$ and $v' = \vec{R'} - \vec{C'}$,

$$T = \begin{bmatrix} u_1' & v_1' & C_1' \\ u_2' & v_2' & C_2' \\ 0 & 0 & 1 \end{bmatrix} \qquad (3.28)$$

This fact is also true conversely: knowing a pair of conjugate diameters of an ellipse, this conjugate diameter pair will define an affine transformation of the form (3.28) which will transform the unit circle into the given ellipse. The transformation in (3.28) can be combined with the parametric equation in (3.26) to produce the parametric equation of the ellipse, that is (by using a vector equation to simplify the formulae):

$$\phi(t) = cos\,(t)u' + sin\,(t)v' + C' \quad (0 < t \le 2\pi) \tag{3.29}$$

This formula is the one which has been adopted by the CGI Standard Proposal to describe an ellipse. Clearly, formula (3.29) is the kind of parametric equation described in the introduction; the characteristic set of points consists of the centre and the endpoints of two conjugate radii.

Knowing the characteristic points, the matrix (3.28) can be described easily. Furthermore, by applying formula (2.44), the matrix of the ellipse can also be established by using

$$(T^{-1})^T A (T^{-1}) \tag{3.30}$$

If a curve in space has to be described instead of a planar one, the same method can be applied; instead of A, one should take A_c, that is:

$$A_c = \begin{bmatrix} 1 & 0 & 0 & 0 \\ 0 & 1 & 0 & 0 \\ 0 & 0 & 0 & 0 \\ 0 & 0 & 0 & -1 \end{bmatrix} \tag{3.31}$$

For the three dimensional counterpart of T, an additional vector is also necessary, namely $w' = u' \times v' - C'$ (that is, the normal of the plane of the curve should be used for the third difference vector). The transformation T should be such that the normal of the plane containing the curve should be transformed onto such a normal vector again; that is the vector $(0, 0, 1)$ should be transformed to $u' \times v'$. The resulting matrix (which replaces (3.28) for 3D) is:

$$T_c = \begin{bmatrix} u_1' & v_1' & w_1' & C_1' \\ u_2' & v_2' & w_2' & C_2' \\ u_3' & v_3' & w_3' & C_3' \\ 0 & 0 & 0 & 1 \end{bmatrix} \tag{3.32}$$

Using A_c and T_c, the matrix of a generalised cylinder is described in space, which describes the ellipse in space; however, the vector equation (3.29) will automatically remain valid.

3.3.2.2. Hyperbolae

The case of the hyperbola is quite similar to that of an ellipse; only the resulting formulae will be a little bit more complicated. The starting point is again a simple hyperbola, which is the one described by the equation $x^2-y^2 = 1$ (figure 3.16).

A parametric equation may also be given for that curve:

$$(1/\cos(t),\tan(t))^T \quad (0 < t \leq 2\pi) \tag{3.33}$$

This parametric equation is given for example in [Penn86] without proof. However, it is not particularly known and it might be interesting to make a small detour to see how the validity of this equation may be proven. Let M be the *projective* (and non–affine) transformation for which the following relations hold (described in the homogeneous coordinates of $I\!PE^2$ generated by the Cartesian ones):

$$
\begin{aligned}
[(1,0,1)] &\rightarrow [(1,0,1)] \\
[(0,1,1)] &\rightarrow [(1,1,0)] \\
[(-1,0,1)] &\rightarrow [(-1,0,1)] \\
[(0,-1,1)] &\rightarrow [(1,-1,0)]
\end{aligned}
\tag{3.34}
$$

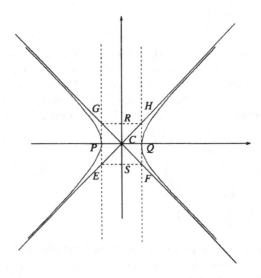

Figure 3.16.

M transforms a unit circle onto the hyperbola of figure 3.16, by transforming the point R (of figure 3.14) onto the ideal point of the hyperbola asymptote $C\mathrm{v}H$, and the point S of figure 3.14 onto the ideal point of the other hyperbola asymptote. These relations might be checked easily by applying the matrix–vector multiplication. The image of the circle will be the hyperbola of figure 3.16. The matrix of M can also be established without too much difficulty:

$$M = \begin{bmatrix} 0 & 0 & 1 \\ 0 & 1 & 0 \\ 1 & 0 & 0 \end{bmatrix} \tag{3.35}$$

will do. Applying this matrix to the equation (3.26), the result is (3.33) (the singularities correspond to the ideal points of the curve). The matrix of this basic hyperbola is again very simple, namely:

$$A = \begin{bmatrix} 1 & 0 & 0 \\ 0 & -1 & 0 \\ 0 & 0 & -1 \end{bmatrix} \tag{3.36}$$

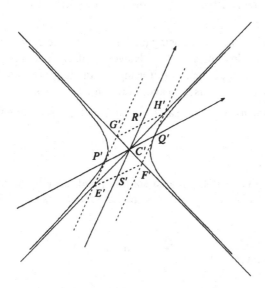

Figure 3.17.

Here again, the two main axes form a conjugate diameter pair, just as in the case of a circle. The significant difference is the exact geometrical description of the points R and S of figure 3.16. As one can see from the figure, the points E,F,G,H are the intersection points of the two asymptotes and the tangents of the curve at P and Q respectively. The asymptotes themselves are also tangential lines; in fact, they are the two tangents at the two ideal points of the conic. Finally, the points R and S may be generated as the intersection points of the Y axis and the lines $G \vee H$ and $E \vee F$ respectively.

An affine transformation (in fact, all projective transformations) transforms tangents into tangents. In other words, an arbitrary affine transformation will transform the configuration of figure 3.16 into a configuration like the one of figure 3.17; C will be transformed into C', P into P' etc. The important fact is that the construction described for R and S uses affine invariant properties only, that is the image of S will be S' (respectively, R will be transformed into R'). Defining therefore an affine transformation which transforms figure 3.16 into figure 3.17 by $C \rightarrow C'$, $Q \rightarrow Q'$ and $R \rightarrow R'$ (whose matrix will be given by formula (3.28) again!), the parametric equation of the hyperbola is:

$$\phi(t) = \frac{1}{cos\,(t)} u' + tan\,(t)v' + C' \quad (0 < t \leq 2\pi) \tag{3.37}$$

(where u' and v' have the same meaning as in case of an ellipse).

Knowing the centre and the points marked by P,Q,R,S in the figure, the parametric equation of the hyperbola can be reconstructed; in other words, this set of points might be considered as being the characteristic set of points for a hyperbola.

Care should be taken when using (3.37) to render the curve; the points tend to infinity, that is overflow may occur. However, these overflows correspond to "infinite points", that is an upper limit using the the largest machine representable floating point number can be used to avoid run–time problems[t].

The three dimensional case can be treated analogously to the two dimensional one.

3.3.2.3. Parabolae

Again, a simple parabola is taken as a starting point, namely the one described by the equation $x^2 = y$ (figure 3.18). A parametric equation can also be given for that curve:

$$\left[\frac{cos\,(t)}{1 - sin\,(t)} , \frac{1 + sin\,(t)}{1 - sin\,(t)} \right]^T \quad (0 < t \leq 2\pi) \tag{3.38}$$

The singularity corresponds to the ideal point of the curve. The interval for the parameter t might be changed; in formula (3.38) the approximation will begin at $(1,1)$, will "go around" through $(-1,1)$ and $(0,0)$. Choosing (for example) the interval $-\pi/2 \leq t \leq 3\pi/2$ would give a more symmetrical arrangement.

Just as in the case of hyperbolae, the equation is known (and can be found in [Penn86] again without proof). The validity of it can also be shown with the same

---

[t] An alternative and somewhat better known equation for the hyperbola would have been:

$$\phi(t) = ch\,(t)u' \pm sh\,(t)v' + C'$$

using the so called hyperbolic cosine and hyperbolic sine functions. However, in this case the parameter t is defined on the infinite interval $0 \leq t \leq +\infty$, which would be computationally unstable.

technique as for the hyperbolae: let M be the *projective* (and non–affine) transformation for which the following relations hold (described in the homogeneous coordinates of $I\!PE^2$ generated by the Cartesian ones):

$$
\begin{aligned}
[(1,0,1)] &\rightarrow [(1,1,1)] \\
[(0,1,1)] &\rightarrow [(0,1,0)] \\
[(-1,0,1)] &\rightarrow [(-1,1,1)] \\
[(0,-1,1)] &\rightarrow [(0,0,1)]
\end{aligned}
\tag{3.39}
$$

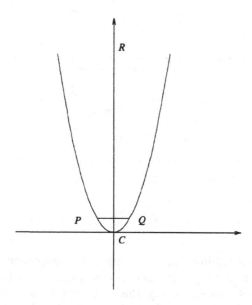

Figure 3.18.

M transforms a unit circle onto the parabola of figure 3.18, by transforming the point R (of figure 3.14) onto the ideal point of the parabola. The image of the circle will be the parabola of figure 3.18. The matrix of M can also be established without too much difficulty:

$$
M = \begin{bmatrix} 1 & 0 & 0 \\ 0 & 1 & 1 \\ 0 & -1 & 1 \end{bmatrix}
\tag{3.40}
$$

will do. Applying this matrix to equation (3.26), the result is (3.38). (the singularity corresponds to the ideal point of the curve). The matrix of the curve is again very simple, namely:

$$
A = \begin{bmatrix} 1 & 0 & 0 \\ 0 & 0 & -1/2 \\ 0 & -1/2 & 0 \end{bmatrix}
\tag{3.41}
$$

The Y axis of the parabola is a diameter; in fact, the centre of the parabola is the (only) ideal point of the curve. The line PQ is of course *not* a diameter in this case (in contrast to ellipses and hyperbolae); it is, however, conjugate to the Y axis (in figure 3.18, the pole of the Y axis is the intersection point of the ideal line and the X axis, which is the direction of this latter one; this coincides with the direction of PvQ). It is also known that the point R will be the middle point of the line segment PvQ.

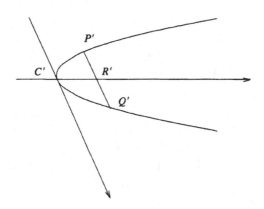

Figure 3.19.

An affine transformation keeps conjugation and keeps also the property of being the middle point of a line segment (this corresponds to the fact that an affine transformation keeps the division ratio). This means that transforming the parabola of figure 3.18 to the one of figure 3.19 would result in the points P',Q',R' and C', where the lines $P'vQ'$ and the axis running through C' (which is the same as the line $C'vR'$) will be conjugate to one another. Furthermore, R' will be the middle point of the line segment $P'Q'$.

The same methods as before can be therefore applied. In this case, however, the meaning of u' and v' is different; indeed, $u' = \overrightarrow{Q'} - \overrightarrow{R'}$ and $v' = \overrightarrow{R'} - \overrightarrow{C'}$. Using these definitions the transformation described in (3.28) will transform the parabola of figure 3.18 into the parabola of figure 3.19. The characteristic points are therefore P',R',Q' and C'; the parametric equation is:

$$\phi(t) = \frac{cos(t)}{1 - sin(t)}u' + \frac{1 + sin(t)}{1 - sin(t)}v' + C' \tag{3.42}$$

$$(-\pi/2 < t \le 3\pi/2)$$

The three dimensional case can be treated analogously to the two dimensional one.

4. Modelling Clip

4.1. Problem Description

All graphics systems include some form of clipping, that is the specification of an area on the plane or in space (depending on the dimension of the system) to which the visible output is confined. In older systems or functional specifications (Core, GKS, GKS–3D, CGI etc.) this clipping area was defined to be a rectangle or a cube; furthermore, the vertices were also required to be parallel to the main axes of the Cartesian coordinate system. In some cases, the user had the option to define more than one such clipping rectangle or cube (for example in case of GKS–3D one may define the Normalisation Clipping Volume, the Projection Viewport and the Workstation Window) and the actual implementation had the possibility to merge these volumes as far as practicable.

There is an extensive literature available on how to perform this clip for line segments, convex and non-convex polygons etc. The usual approach makes use of the convexity of the clipping body. Primitives are clipped against the half–spaces describing the volume itself in a pipeline fashion: they are clipped against the first half–space, the output of this step is clipped against the second one and so on. The advantage of this approach is that it helps a really parallel implementation by putting one clipping step into a separate process or processor. This approach goes back to as far as 1974 ([Suth74]) but there have been no really widespread new approaches ever since. The reader may find a description of the different methods in most of the usual and already cited textbooks or tutorials.

The clipping problem itself, though being relatively simple from a mathematical point of view (at least for simple primitives), plays a very important role in realising efficient implementations. Indeed, all output primitives have to be clipped at some point in the output pipeline. This means that even if the clipping step is simple, an inefficient realisation of it may slow down an otherwise very well implemented graphics system. It is no surprise that clipping against a rectangle was one of the first graphics functions (apart from the drawing functions themselves, of course) to be implemented on silicon.

The appearance of PHIGS as a widespread standard proposal has created a new clipping problem or, to be more exact, has raised the need for a more general form of clipping. Traditionally, clipping was designed primarily to "cut" some part of the image generated for an object when placing it onto the screen (or the plotter) rather than to cut some part of the object itself in a more arbitrary way. The introduction of arbitrary clipping planes allows a user to select parts of a model for display. These planes define a more general volume in space (or, in 2D, an area of a plane, although PHIGS being a 3D standard, the specifications are all for the 3D case) which serves as a clipping volume and which is independent of the viewing step. This is known as a *modelling clip*.

With a modelling clip the user has the possibility of defining an arbitrary number of half–spaces the intersection of which forms a convex (not necessarily bounded) body; this body is called the modelling clipping volume. All primitives are trimmed to the interior of this volume and only those parts of the primitives inside the volume are potentially visible on the screen. Furthermore, when traversing the geometric structures to be rendered on screen, a new current clipping volume can be created by combining the old clipping volume and a modelling clipping volume to yield a new current clipping volume. In this case "combining" is defined as a set–theoretic function (ie intersection, difference, union) applied to the incoming clipping volume and the convex body. Thus, the current clipping body can be a complicated, non convex and not even connected area in space.

There are basically two problems arising from the modelling clip. First, the possibility of using set–theoretic combination for the construction of a current clipping volume leads to the use of *shielding*; shielding is the opposite of clipping in the sense that the part of a primitive which is *not* in the clipping volume is to be accepted. The effect of this is to augment the number of output primitives, making the efficiency of the whole graphics pipeline even more dependent on the clipping step itself. Furthermore, it is also a non–trivial programming problem to keep track of all the combinations of convex bodies which comprise the current clipping volume as well as their effects on the output primitives.

The difficulties involved in implementing the full modelling clip as described above were recognised by the ISO/IEC SC21 Working Group working on the PHIGS proposal. While in some earlier versions of PHIGS (up to 1987) all the 16 set–theoretic combinations were mandatory, this is not the case any more in the actual version ([ISO89]) where only the intersection is demanded; the choice of supporting all possible combinations is left to the implementor. It must be remarked, however, that having the full specification in hand *is* very useful indeed: a number of necessary effects cannot be created otherwise. O'Bara and Abi–Ezzi give some nice examples in [OBar89]. In other words, a really prestigious implementation should probably choose to implement all of the possible combinations[t].

The programming problem described above has been solved elegantly by O'Bara and Abi–Ezzi by using the so–called CCV–Filters (see [OBar89]). As this filtering problem does not involve projective geometrical aspects, it is not described here; the reader is referred to the article itself. The important consequence of this result is that one can concentrate on the geometric clipping/shielding against convex bodies only; the CCV–Filters provide a solution for the final modelling clip itself based on the output of the individual convex body clips (and the way these CCV–Filters work does not involve geometrical calculations any more).

The other problem concerning modelling clip is as follows. As far as the effective clipping is concerned, at first glance it does not seem to be particularly difficult: the fact of having an intersection of half–spaces suggests that clipping may be performed by a series of clips against half–spaces: the output of one step should

---

[t]In fact, very few commercial PHIGS implementations have implemented the modelling clip even in its simplest version. This also indicates the difficulties involved.

be used as an input for the next one. In PHIGS, however, the real problem arises as a result of the fact that the modelling clip is to be performed after the so called (composite) modelling transformation, which is a projective transformation (that is, defined by a 4×4 matrix). Using the notations of PHIGS, the modelling transformation transforms the *Modelling Coordinate System (MC)* into the *World Coordinate Space (WC)*.

The data defining a half–space (that is a point and a normal vector) are defined by the application program in MC. The half–space itself is to be transformed by the modelling transformation to define an appropriate point–set in WC, and the modelling clip has to be performed against this set. This is how the process is described and defined in the PHIGS document. The half–spaces have to be transformed by a projective and (in general) *non-affine* transformation, and it is the use of such transformations which may lead to mathematical problems. Indeed, the notion of half–space has no meaning any more in projective space; what will be the image of a half–space? Clearly, the problem of W–wraparound, described earlier for line segments, arises very seriously in this case as well.

It is a natural idea to think that the whole modelling clip might be performed in MC rather than in WC; this would avoid problems. Unfortunately, this is not possible. The reason for this restriction lies in the structure model of PHIGS. In fact, the current modelling transformation and the current modelling clip volume itself is subject to eventual changes as a result of structure traversal. And, as stated in PHIGS:

> "... when one of these structure elements[†] is encountered during structure traversal, each half–space specified is transformed by the current composite modelling transformation. The resulting clipping volume is not affected by subsequent transformation encountered during structure traversal." [ISO89].

That is, if the current transformation is changed during structure traversal, the transformation used to transform the output primitives and the one having been used for the (modelling) clipping volume are *not* the same any more. In other words, if the clipping were performed in MC, the clipping volume itself should be transformed from one environment to the other, which would lead to a projective transformation again.

Clearly, the real source of the problems is the fact that the modelling transformation is allowed to be a non–affine transformation as well. If it were restricted to be an affine transformation (that is a combination of rotation, translation, scaling and shearing), no problem would occur and the modelling clip could be performed on the "top" of the pipeline prior to any kind of transformation (although, as presented later, this is not necessarily the optimal solution!) This question had been at the source of a number of discussions when describing the PHIGS Functional Specification and has also been misunderstood by a number of authors (for example,

---

[†]That is structure elements defining half–spaces.

the otherwise very important paper of O'Bara and Abi–Ezzi cited above fails to handle this problem correctly). The reason why the full specification has been finally adopted was that there are some effects which cannot be rendered properly without the use of a fully projective transformation: a typical example is when the user wants to model a projective environment![t] In spite of that it remains valid that the modelling transformation will most probably be an affine transformation instead of a fully projective one.

There is however a more subtle reason why the use of a fully projective transformation is advantageous in all cases even if the modelling transformations themselves were restricted to be affine. If there was a way to perform the modelling clip properly, that is *after* a general projective transformation, this would enable the possibility to *merge* the modelling transformation with the viewing itself, resulting therefore in one single transformation in the output pipeline (this will be detailed later). That is, instead of using two transformations (in other words two times a vector–matrix multiplication) one would suffice. Additionally, performing clipping before the transformation would have a number of side–effects. For example, as mentioned before, clipping (and shielding) has the disagreeable effect of eventually augmenting the number of points and output primitives. In other cases clipping of the primitives might not be very simple: for example, if an implementation chooses to approximate conics after the linear part of the transformation, as described in the previous chapter, it becomes necessary to perform the clipping after the transformation (that is after the generation of the approximating polygon). Indeed, clipping of a conic is not an easy operation. In other words it does make the output pipeline implementation more effective if the modelling clip is done after the transformation rather than before (provided of course that the clipping does not become too complicated) even if the modelling transformation itself may be considered as being affine.

With these facts in mind, the modelling clip problem can be described in relation to the output pipeline in a somewhat more precise and general way (additionally, this formulation may be used for environments which are different from PHIGS, like PEX [Clif88], [ISO88b]):

- A list of half–spaces is defined in the three dimensional Euclidean space;
- the half–spaces (or the intersection of the half–spaces) are transformed by a projective, not necessarily affine transformation to form a clipping volume;
- all output primitives should be clipped after a transformation (which is not necessarily the one used for the transformation of the clipping volume) against the (transformed) clipping volume.

Figure 4.1 illustrates that there is a real problem to deal with. The figure shows one possible effect of the W-wraparound problem as far as the image of a half–space (on the figure a half–plane) is concerned. A line l on the plane Π is transformed onto the

---

[t] For example, the user might want to describe an object as well as the *projected* image of an object, which must be done independently of the viewing itself.

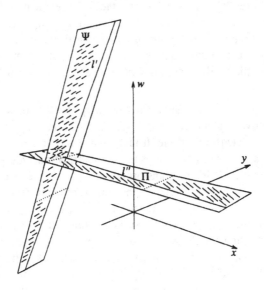

Figure 4.1.

line l' of the plane Ψ. The corresponding half–plane will still be transformed onto a half–plane of Ψ. However, when performing the projective division, the image of l' will be l'' and the image of the half–plane will be a more complicated area on Π. (To avoid confusion, the line l is not in figure 4.1). The separating line between the two areas on Π (shown in the figure) is the one which is cut by a plane parallel to Ψ and crossing the origin.

As mentioned before, though contained in the different versions of the PHIGS specification, no actual implementation existed for the modelling clip for a long time. It was in the paper of Herman and Reviczky ([Herm88]) that the mathematical and algorithmic problems generated by the modelling clip were identified for the first time and this article provided also the first known solution[†]. Another approach has been proposed one year later by Krammer in his already cited paper [Kram89] and, finally, the paper of Hübl and Herman ([Hübl90]) has provided with additional details. This last paper may be considered as giving a final mathematical description of the modelling clip as a whole. What will be presented in this chapter is the mix-ture of these three papers; the author knows of no other published results on the mathematics of the modelling clip.

The solution of Krammer is very simple once the underlying projective geometrical principles are understood. As described in chapter 2, the notion of a *conic sector* might be considered as the generalisation of a half–space. A convex

---

[†] Although, unfortunately, it fails to handle properly the fully general case.

body, being the intersection of a certain number of half–spaces, is also the intersection of a number of conic sectors: indeed, each half–space can be considered as a conic sector with one of the generating planes being the ideal plane. Consequently, the image of the convex body is the intersection of all these conic sectors. The modelling clip can therefore be performed after the transformation by performing a clip against these conic sectors.

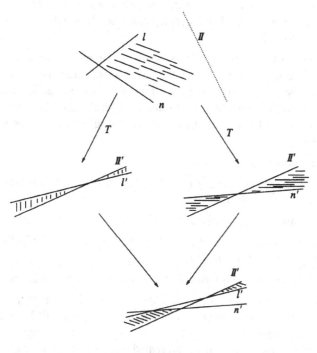

Figure 4.2.

This is probably the cleanest way to describe the modelling clip effect with the terminology of projective geometry and it gives a clear means to implement the clipping itself in a graphics system (this has been done in the IXPHIGS implementation described in [Görö88]). Figure 4.2 shows however the drawbacks of this approach (in 2D). The lines l and n generate a convex (unbounded) area. Using II, the two half–planes are to be considered as conic sectors. The transformation T transforms the ideal line onto II', which is considered now to be an affine line. l is mapped onto l' and n is mapped onto n'. The conic sectors are now "real" conic sectors and not half–spaces; consequently, the intersection of the conic sector $l'\mathit{II}'$ and $n'\mathit{II}'$ might generate the two unbounded areas shown in the figure. In other words, the conic sector approach may augment the number of clips to be considered and this might slow down the modelling clip substantially (on the figure, the number of half–space clips becomes 5 instead of the original 2). It is therefore necessary to try to reduce this number whenever possible. Unfortunately, this "duplicating" effect cannot be avoided when a fully general solution is sought. But, when trying for example to perform the modelling clip in 4D, some important special cases can be described when this problem does not occur.

4.2. Solutions in 4D

The background for the 4D solution is the same as used throughout chapter 3: try to transfer the problem onto the four dimensional Euclidean space. As seen before, this approach has the advantage of still keeping the Euclidean nature and properties of all the objects involved.

A special case examined first is when the transformations applied to the current modelling clipping body and the one transforming the output primitives themselves are the same. This is the most probable case in practice: the modelling clipping body is usually used to "cut" some parts of the primitive to be visualised; the most natural way of describing this "cutting" volume is to use the same modelling transformation.

The method is then as follows:

i)   transform all half–spaces by the linear part of the transformation (that is by the matrix–vector multiplication);

ii)   transform all output primitives by the linear part of the transformation;

iii)   perform a clip against the images of the half–spaces in a "pipeline" fashion; this step has to be done in $\mathbb{R}^4$;

iv)   perform the projective division on the (clipped) output primitives.

The real question is of course how to do steps i) and iii). For that purpose, it suffices to concentrate on what the image of one single half–space will be and how these steps may be performed in this case, since the modelling clip body is built up of such half–spaces. The ideas will be presented with figures showing, as usual, the two dimensional analogy.

Points lying in one half–space (or half–plane) are characterised by the boundary plane/line. This boundary is usually specified by giving its Euclidean normal n, pointing toward the selected area, and one point of the boundary itself. Using the notations of figure 4.3, the boundary will be denoted by l and the half–space (resp. half–plane) by $\Pi_{l,n}$ or simply $\Pi_l{}^\dagger$.

l being a plane/line in projective sense as well, it has homogeneous coordinates u; furthermore, these homogeneous coordinates determine an *Euclidean* plane (resp. line) in $\mathbb{R}^4/\mathbb{R}^3$ by using the homogeneous coordinates as a vector parallel to the normal of this plane. This plane (denoted by Ω on the figure, see also chapter 2.5) also determines a half–space in $\mathbb{R}^4/\mathbb{R}^3$ denoted by $\mathbb{R}^n{}_\Omega$. Clearly:

[†] In fact, $\mathbb{R}^3$, should be used but because of using the straight model of $\mathbb{P}E^2$ on all figures it is better to keep to this notation.

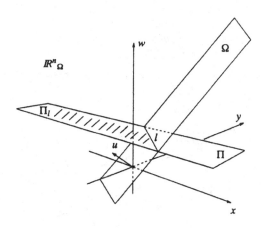

Figure 4.3.

$$l = \Pi \wedge \Omega \qquad (4.1)$$

and, furthermore:

$$\Pi_l = \Pi \cap I\!R^n{}_\Omega \qquad (4.2)$$

Applying the linear part of the projective transformation, this linear part will be an affine transformation of $I\!R^4/I\!R^3$. An affine transformation will turn a half-space/half-plane into a half-space or a half-plane respectively. As usual Π, which is the straight model of $I\!PE^3$ or $I\!PE^2$, will be mapped onto Ψ; at the same time, the plane Ω will be mapped onto the plane Θ.

Θ will still cross the origin. By denoting the image of l by l', the following relations also hold (see figure 4.4; neither Ω nor l are shown in the figure to avoid confusion)):

$$l' = \Psi \wedge \Theta \qquad (4.3)$$

and

$$\Psi_{l'} = \Psi \cap I\!R^n{}_\Theta \qquad (4.4)$$

The original goal of the modelling clip is to clip the primitives (which are all part of Π) against the half-space $\Pi_{l,n}$. Obviously, Π_l is mapped onto $\Psi_{l'}$; furthermore, relations (4.2) and (4.4) mean that it is possible to postpone this clipping. The primitives can be transformed by the matrix–vector multiplication to result in some geometric primitives in $I\!R^4$ ($I\!R^3$ on the figures) and then clip them against $I\!R^n{}_\Theta$. This latter step is nothing other than a normal clip using a hyperspace as a boundary; as described in relation to the W–clip, it does not create any particular complications to

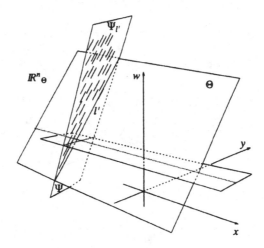

Figure 4.4.

perform this clip in four dimensions.

To perform the above described steps numerically, a vector which is parallel to the normal vector of Θ should be found in $I\!\!R^4$. There are several ways of doing this. The approach proposed in [Herm88] was to choose three non–collinear points on l, plus a point E in Π_l, transform these points and use the general form of the outer product (already used in case of the W–clip optimisation) to determine a vector perpendicular to Θ. The image of E should then determine the sign of the outer product. In reaction to this paper, Zachrisen has proposed an alternative and more elegant way of calculating this vector in [Zach89]. His approach is as follows. Suppose that T is the projective transformation in use, n is the (three dimensional) normal vector describing l and p is a point of l. The points of $\Pi_{l,n}$ can be described as follows:

$$\Pi_{l,n} = \{x \in I\!\!R^3 : \sum_{i=1}^{3} x_i n_i + \alpha > 0\} \tag{4.5}$$

where $\alpha = -p^T n$. This also means, however, that in homogeneous coordinates the vector:

$$u = [(n_1, n_2, n_3, \alpha)] \tag{4.6}$$

will be the homogeneous coordinates of l or, in other words, this will be a vector parallel to the normal vector of Ω.

Θ is the image of Ω after the linear transformation T. It is a well–known formula of linear algebra that the image can be characterised by the vector:

$$u' = (T^{-1})^T u \qquad\qquad (4.7)$$

In other words, (4.7) gives a vector parallel to the normal vector of Θ.

In the special case described here the modelling clip can be solved without duplicating the number of necessary clips at the price of performing the clipping steps themselves in 4D. As seen before, however, all usual (3D) clipping algorithms can be generalised for 4D as well without any additional complication; although the calculations must then be performed for a fourth coordinate component as well, this is still less than performing yet another clipping step for each element in 3D.

The reason why this approach cannot be extended to a general case is that the Ψ–s (to use the usual notations) become different if the transformations are changed. If no statement can be given about the mutual relationships of the transformation used for the current clipping body and the one used for the primitives themselves, it becomes very complicated to describe what the mutual spatial relationships of the two Ψ–s can be.

In practical circumstances however the situation is not so bad. Remember that the output pipeline of a typical 3D system (like PHIGS) is schematically as follows:

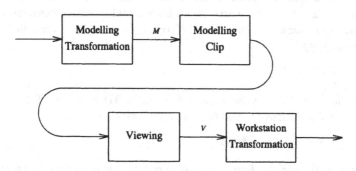

and, in fact, the transformation which is really in use is $T=VM$, that is first the modelling and then the viewing transformation respectively. The real advantage of combining the modelling clip with a general projective transformation is the fact that the output pipeline can then be rearranged as follows:

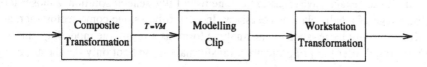

which means, in practice, that only one matrix–vector multiplication has to be used instead of two.

The main problem using the four dimensional approach is that in reality the M–s can change relatively frequently; that is, there might be an M_1 used for the current clipping volume and another M_2 for the primitives themselves. In the case of PHIGS, V can also change (the user may use structure elements, called "set view index" which change the actual view transformation in use). In most of the practical cases, however, the change of this latter transformation is not really frequent; in some 3D packages, like for example Doré ([Arde87]), it is even confined to a group of output primitives (more or less the equivalent of a PHIGS structure). One may therefore suppose that for a large number of objects V is always the same. In other words, the difference between the transformations $T_1 = VM_1$ and $T_2 = VM_2$ is fully described by the differences between M_1 and M_2.

As said before, the specification of PHIGS allows an arbitrary 4x4 matrix for M, that is an arbitrary projective transformation can be used. It is also true, however, that in most of the practical cases M will be *affine*; indeed, the primary use of this transformation is to move and eventually scale the objects to be visualised. Furthermore, going again beyond pure PHIGS and considering other general 3D systems, this restriction might just be part of a functional specification. As an example: the output pipeline of GKS–3D may be regarded as a special case of the one of PHIGS: the segment and normalisation transformations of GKS–3D are special (and affine!) modelling transformations and the normalisation clip may be regarded as a special modelling clip.

Whether a transformation is affine or not can be detected very easily by inspecting the last row of its matrix. Apart from a multiplicative constant, if affine, this last row will always take the form of $(0,0,0,1)$. It is however clear that for such a transformation M:

$$M(\Pi) = \Pi \tag{4.8}$$

so that the last coordinate value of any point of $M(\Pi)$ is still 1. Taking the two modelling transformations M_1 and M_2, the following also holds:

$$\Psi = VM_1(\Pi) = VM_2(\Pi) \tag{4.9}$$

In other words, the use of a four dimensional clip for the modelling clip is still valid in this case. Clearly, this means that the four dimensional clip can be applied for most of the practical cases, even if it fails to cover the fully general one.

4.3. A General Solution

In order to see exactly what has to be done for a fully general solution, a closer look at the image of a half–plane is necessary. In what follows, the conic sector approach of Krammer will be re–stated in an analytic form; as a result, the modelling clip will be performed after the *full* projective transformation (and not only the linear part of it).

The main result presented in this section is the following. If $n \in \mathbb{R}^4$ is an arbitrary non–zero vector, it defines a half-space in the Euclidean space $\mathbb{R}^3$ by using the following formula:

$$\{n\} = \{x \in \mathbb{R}^3 : \sum_{i=1}^{3} n_i x_i + n_4 > 0\} \tag{4.10}$$

The notation $\{n_1, n_2\}$ will also be used to denote $\{n_1\} \cap \{n_2\}$. The following theorem will be proved:

Theorem 4.1. Let M be a projective transformation, u the homogeneous vector describing a half–plane and $v = (0,0,0,1)$ the vector describing the ideal plane. Let u' and v' be the vectors describing the image of these planes, that is (applying formulae (4.7)):

$$u' = (M^{-1})^T u \qquad\qquad (4.11)$$

and

$$v' = (M^{-1})^T v \qquad\qquad (4.12)$$

Then the affine part of the image of the half–space is (see also figure 4.5):

$$\{u', v'\} \cup \{-u', -v'\} \qquad\qquad (4.13)$$

Figure 4.5.

Theorem 4.1 describes, in fact, a conic sector in analytical form. By induction, the following statement is also true:

Theorem 4.2. If a convex body is determined by the set of homogeneous vectors $u_1, \ldots, u_k$, the image of the convex body under the effect of a projective transformation will be:

$$\{u'_1, \ldots, u'_k, v'\} \cup \{-u'_1, \ldots, -u'_k, -v'\} \qquad\qquad (4.14)$$

This result is the analytic reformulation of the conic sector intersection method of Krammer; however, it gives more insight into the expected result. Indeed, theorem 4.2 means that the image of a convex body will be the union of two (disjoint) convex bodies described by (4.14). This result has been proven first in [Hübl90].

Theorem 4.2 might be surprising at first glance. Knowing that a half–space might be mapped onto a real conic sector duplicating the convex areas, one might expect that the final outcome for a whole convex body becomes more complicated. It is therefore worth examining a more intuitive picture of this description. Using the

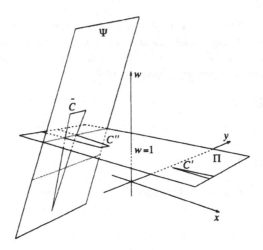

Figure 4.6.

usual straight model, figure 4.6 shows what happens to a simple convex body (in this case a triangle). The original convex body (not shown on the picture) is transformed onto $\bar{C}$; $\bar{C}$ being the affine image of C, it is a convex body on Ψ. By performing the projective division, this convex body is projected onto C' and C''; the effect is the usual W–wraparound problem in case of a whole convex body.

Figure 4.7 shows also that one of the two convex bodies in 4.2 can also be empty: indeed, if $\bar{C}$ is fully on the $w>0$ or the $w<0$ side of Ψ, this will be the case.

The proof of the theorems is as follows. If $p \in I\!PE^3$ is a point of the half–space defined by u and the coordinates of p are such that $p_4=1$, the following holds:

$$p^T u > 0 \tag{4.15}$$

Equation (4.7) also means that if $p'=Mp$, the corresponding *four dimensional* vectors p' and u' fulfil the following relation as well:

$$p'^T u' = \sum_{i=1}^{4} p'_i u'_i > 0 \tag{4.16}$$

This is, in fact, what has been used in the previous section for the four dimensional approach.

Let consider first $p'_4>0$. In this case, formula (4.16) can be divided by p'_4 leading to:

$$\sum_{i=1}^{3} \frac{p'_i}{p'_4} u'_i + u'_4 > 0 \tag{4.17}$$

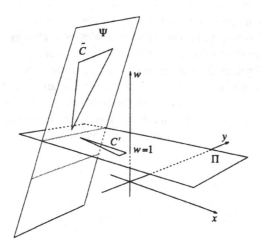

Figure 4.7.

This means, in other words, that the *affine image* of p will become an element of $\{u'\}$.

If $p'_4 < 0$, the final affine image will be an element of $\{-u'\}$. Finally, p being an affine point originally, the same line of arguments can be followed for v instead of u. Combining the formulae leads to the desired results. ∎

It is now clear how the modelling clip can be implemented effectively. The two convex bodies described in theorem 4.2 give the image of the modelling clip body; all clipping against these bodies can therefore be done *after* the full projective transformation. It is, of course, a natural and also important question to find out whether both bodies exist or not or whether the clip against both of them is necessary or not.

First of all, if M is affine, either v' or $-v'$ will be equal to $(0,0,0,-1)$. Consequently, the corresponding convex body will be empty.

An obvious check a system may and should make is similar to the one used for the optimisation of the W–clip: it should be checked to see if the workstation clipping cube is disjoint with one of the two convex bodies. This can be done, obviously, by checking whether the workstation clipping cube is fully within one of the half–spaces listed in theorem 4.2. If yes, the *other* clipping body (if it exists at all) can be disregarded. The best candidate for that purpose is v' or $-v'$: indeed, the boundary plane of these half–spaces is the image of the ideal plane. On the other hand, it has been shown that in the case of a synthetic camera model, the image of the ideal plane will be disjoint of the workstation clipping cube; in other words, the cube will be disjoint with either $\{v'\}$ or $\{-v'\}$. The synthetic camera model being one of the most important viewing models in practice, this fact is very significant.

It is a question of debate whether the four dimensional clip or the one described in the present section is more advantageous (besides that this latter one covers the problem in full generality). There does not seem to be much difference in the computational demands of the respective methods, although no actual comparison has been done; the choice seems to be rather a question of taste. It is also dependent on the features used in general: if an implementation chooses to use the general approach described in chapter 3 to handle cell array, STROKE characters etc, it becomes attractive to incorporate the modelling clip as well into the four dimensional environment. Credit should be given to the four dimensional approach anyway, being historically the first attempt to give a solution for the problems of modelling clip.

5. Projective Algorithms

5.1. Introduction

The algorithms and methods which will be presented in this chapter are radically different from those of the previous ones. Whereas the basic idea behind all methods up to now was to try to reformulate the problems arising for four dimensional space (with the exception of the modelling clip body description), in this chapter the problems will be described in purely projective terms. In other words, in chapters 2 and 3 the idea was to get "back" into an Euclidean environment via the use of four dimensional geometry (and hence making use of all traditional methods already in use in computer graphics), whereas in the present chapter the problems will be described in purely projective geometrical terms. This is the reason why all these algorithms may be called "projective" algorithms.

There are several reasons why this is of interest. First of all, the methods which will be described will give a *projective invariant* formulation of some primitives. This also means that the implementor of a 3D graphics system might have a choice when implementing for example cell array: instead of generating the cells in 4D (as described previously), the projective invariant description of the cell array may be also used, which means that a number of points are transformed by the *full* projective transformation, and, as a next step and making use of these (transformed) points, the full cell array is generated directly on the projective plane. This approach is similar to what has been done for the description of conics in chapter 3; the difference is that instead of affine invariance, projective invariance is sought.

The advantage of using this approach might be algorithmic; in the 4D approach there are three additional divisions to be done on all generated points (eg on the polynomial approximation of a conic). In some environments, however, division is not cheap at all and it might therefore be of interest to reduce their number. Unfortunately, there has been no opportunity to make an effective comparison between the clean four dimensional approaches and the ones described below; whereas the four dimensional algorithms described in chapter 3 have become integral part of a commercial product, the projective invariant methods could be tried out on an experimental basis only.

Speed is however not the only issue. Some problems which might make a projective invariant formulation necessary come from a more theoretical standpoint, which is as follows.

General purpose graphics systems and standards, like GKS, GKS–3D, PHIGS, PHIGS PLUS, CGI or any of their ancestors like GINO–F ([Wood71]), GPGS or Core ([GSPC77] and [GSPC79]), can be classified rigorously as either 2D or 3D systems. The notion of 2D systems means that the function definitions contain graphics output primitives for two dimensions only; lines, markers, polygons etc defined on a plane. These primitives may be fairly complicated, like STROKE precision characters or polygons with patterned interiors. However, the definitions of these primitives reflect the planar nature of graphics: for example, patterns are defined as a regular set of small parallelograms, dividing in turn a larger parallelogram regularly etc.

3D systems usually contain the more or less obvious generalisations of the two dimensional output primitives: polygons, text etc. are defined with three dimensional coordinates together with more complicated primitives like for example B–Spline or NURB[†] curves or surfaces in packages like PHIGS PLUS, HiRasp, Doré etc. Furthermore, these systems contain different forms of viewing facilities; the ultimate goal of these features is to project the three dimensional primitives onto the display screen which is, at present, inherently two dimensional. In the course of this process the system may also perform some kind of hidden surface and/or hidden line removal.

It is relatively straightforward to find application areas which may use a two dimensional package. A number of practical problems are basically planar: business graphics, some areas of presentation graphics and even some CAD applications; electronic design is a good example.

This is not the case for three dimensional systems. There are of course application areas which are inherently 3D, like mechanical design or architecture. However, when realising these systems the implementors may find it more natural to use a two dimensional general graphics system for their own purposes rather than a three dimensional one. This contradiction is the result of the fact that a sophisticated application program (eg a solid modeller) has to create some kind of internal representation of the objects it wants to manipulate; an internal representation which contains, among other things, a whole range of strictly geometrical data. Consequently, it may happen (and, in fact, it does happen), that the application program has the possibility to perform all viewing, hidden line/surface removal with a significantly higher speed than by using a general 3D system. For that purpose, it can make use of the sometimes rather detailed additional information which is within the internal representation. Using a full three dimensional graphics system would mean in this case the duplication of the viewing pipeline; a duplication which might be expensive both in terms of efficiency and price.

The result of this is that such systems might find it more straightforward to use a two dimensional environment instead of a three dimensional one. However, in this case, one faces a disturbing problem: a number of complicated but at the same time very useful features of these systems are not usable any more! In fact, primitives like cell array, polygon filling with patterns, stroke precision characters are defined in such a way that it is impossible to use them to generate the *planar projection* of their three dimensional counterpart. As an example, it is not possible to generate directly a picture like the one on figure 5.1 with a traditional two dimensional system like GKS (texts may be deformed by affine transformations only, namely with the help of the segment transformation). In other words, the application has to solve for example the pattern filling of a polygon by itself, although this is clearly in

---

[†]Non–Uniform Rational B–Spline

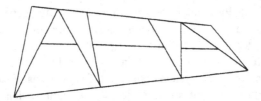

Figure 5.1.

contradiction with the demand of using general purpose graphics systems whenever it is possible to reduce development time and costs.

It might therefore be interesting to see whether it is possible to define an extension of the traditional two dimensional systems so as to cover the geometry of the planar projections of the traditional three dimensional output primitives. This means, in practice, defining projective invariant formulations for those primitives which are the source of the problems.

The problem itself with some possible solutions (described in [Herm89] and in [Herm89a] and presented later in this chapter) has also been raised at the international study group of the ISO/IEC SC24 committee on the so–called New API (Application Programmers' Interface), a study group which is examining a successor to GKS[†]. This is also the reason why the terminology ''2.5D systems'' was used for those enlarged graphics systems (that is 2D systems with projectively invariant specifications) although, unfortunately, this notation proved not to be a very good choice (2.5D has a very different meaning in e.g. CAD literature). The present chapter concentrates exclusively on the purely projective geometrical nature of the problems arising. No attempt is made to give a formal specification of an ''extended'' 2D system; this is the task of the ISO group and is beyond the scope of this study. The mathematical algorithms presented here will however prove that such a specification is feasible and implementable.

It has to be stressed that the usability of all the methods described in some of what follows is underpinned by the general description of the modelling clip, given in the previous chapter. The net result of this description is that the modelling clip (which encapsulates for example the normalisation clip of GKS–3D) can be performed after the full projective transformation. In other words, the results of theorem 4.2 might be considered as a projective invariant description of the modelling clip. This fact is important: to take the example of the cell array, if the clip were to be performed prior to the full transformation, the original parallelogram would be

---

[†]To be precise, in [Herm89] only the starting point of a possible solution for the problems arising in the case of for example cell arrays has been given; the proper handling of ideal points was not presented there. In what follows, a full version will be given.

cut into a more general convex polygon, in which case the methods described would not be usable any more. However, by putting the modelling clip at the end of the pipeline, all internal cells (in the projective environment) can be generated without regard for the clipping and, finally, all generated cells can be clipped individually. Essentially, one can choose freely when to perform clipping and this choice may depend on the primitive in question; this freedom has been made possible by the consequences of theorem 4.2.

In what follows, the output primitives described in chapter 3, that is cell array, pattern filling, STROKE characters (see figure 5.1) as well as conics will be reconsidered; these are typical examples of primitives which are time–consuming to generate and, at the same time, produce very significant distortions when using a projective transformation. Even the W–clip (or any analogous approach) will sometimes be omitted: the aim is to describe these primitives so that singularities are handled automatically. In all cases a *projective invariant* way of specifying the primitive will be given; this specification will be a generalisation of the usual one.

5.2. Regular Subdivisions and Their Images

5.2.1. Regular Subdivision of Lines

The background for the projective distortions can, in a number of cases, be explained by the distortions created on a set of regular subdivision points. This means, that a number of primitives can be described (prior to transformations) by the following data:

- two points P and Q defining a line segment and
- $n-1$ points $A_1, \ldots, A_{n-1}$ on the line segment PQ so that:

$$\overline{A_i A_{i+1}} = \overline{PA_1} = \overline{A_{n-1}Q} = \frac{1}{n}$$

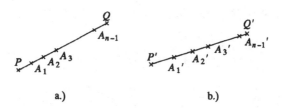

Figure 5.2.

(see also figure 5.2a). Indeed, this is the basic structure used by cell array, patterns etc.

The projective distortion occurs because the points $A_1, \ldots, A_{n-1}$ are not necessarily mapped onto regular subdivision points (see figure 5.2b); this is equivalent to the fact that the division ratio is not a projective invariant factor. The images of these points are not absolutely random, however; in fact a fairly regular structure can be found. This structure will be described in what follows, and will be the clue of most of the results described later.

Although the division ratio is not projective invariant, the value of the double ratio is kept by a projective transformation. It is therefore a natural idea to try to describe the position of the points A_k' (figure 5.2b.)) using this value. What is needed for that purpose is a more explicit formula for the double ratio of points.

Let P, Q, A_k and A_i respectively denote four different affine collinear points (the case when one of these points may be ideal will be dealt with later). To make the presentation simpler, the points A_k and A_i are considered as elements of the line segment PQ (see figure 5.2). The value of the double ratio for these points has been defined (in chapter 2.8) by:

$$(PQA_kA_i) = \frac{(PQA_k)}{(PQA_i)} \tag{5.1}$$

where (PQA_k) (resp. (PQA_i)) denotes the division ratio, defined by:

$$(PQA_k) = \frac{\overline{PA_k}}{\overline{A_kQ}} \tag{5.2}$$

by using directed distances in all formulae. The value of (PQA_kA_i) is denoted by $\alpha_{k,i}$ and it is considered as known.

Formulae (5.1) and (5.2) lead to the following:

$$\alpha_{k,i} = (PQA_k) \frac{1}{\left[\dfrac{\overline{PA_i}}{\overline{A_iQ}}\right]} \tag{5.3}$$

by substituting the value of $\overline{A_iQ}$:

$$\alpha_{k,i} = (PQA_k) \frac{\overline{PQ} - \overline{PA_i}}{\overline{PA_i}} \tag{5.4}$$

that is:

$$\alpha_{k,i} \overline{PA_i} = (PQA_k)\overline{PQ} - (PQA_k)\overline{PA_i} \tag{5.5}$$

which leads to a final expression:

$$\overline{PA_i} = \frac{(PQA_k)}{\alpha_{k,i} + (PQA_k)} \overline{PQ} \tag{5.6}$$

This simple equation is the clue to the further results; it shows that if the value of the double ratio is known from some external source, the point A_i may be determined on the line with the help of the other point A_k and the (directed) distance of the latter from P.

In case the points A_k and A_i represent regular subdivision points of the line segment PQ (like on figure 5.2), the value of the directed distances and of the division ratio can be calculated easily. Indeed, if there are n subdivisions, then

$$(PQA_i) = \frac{\overline{PA_i}}{\overline{A_iQ}} = \frac{\dfrac{i}{n}}{\dfrac{n-i}{n}} = \frac{i}{n-i} \tag{5.7}$$

in other words,

$$\alpha_{k,i} = (PQA_kA_i) = \frac{\dfrac{k}{n-k}}{\dfrac{i}{n-i}} = \frac{k(n-i)}{i(n-k)} \tag{5.8}$$

If the points P, Q, A_i, $i = 1,\ldots,n-1$ are transformed by a projective transformation, mapping the point P onto P', Q onto Q' etc, the image points A_i'-s will not remain regular subdivision points. However, if the points are all on the line segment $P'Q'$, the considerations leading to formula (5.6) remain valid and, furthermore:

$$(P'Q'A_k'A_i') = (PQA_kA_i) = \alpha_{k,i} \tag{5.9}$$

which is one of the basic theorems on the mutual relationship of projective transformation and the double ratio. This means, in other words, that formula (5.8) can be used to replace the value of $\alpha_{k,i}$ in formula (5.6):

$$\overline{P'A_i'} = \frac{(P'Q'A_k')}{\alpha_{k,i} + (P'Q'A_k')}\overline{P'Q'} \tag{5.10}$$

It makes the formulae a little bit simpler if $k = 1$; indeed, in this case (using α_i to denote $\alpha_{1,i}$):

$$\alpha_i = \frac{n-i}{i(n-1)} \tag{5.11}$$

It is important to note that the values of α_i (or, equivalently, $\alpha_{k,i}$) are independent of the effective geometrical position of the points P, P', A_i, A_i' etc. In other words, if the value of n is known, these values can be calculated in advance (this might be very important in practice).

It is also easily verifiable that if

$$(P'Q'A_1') = \frac{1}{n-1} \tag{5.12}$$

(that is the point A_1' corresponds to the first regular subdivision point), all other points will automatically be regular, that is formula (5.10) reduces to:

$$\overline{P'A_i'} = \frac{i}{n}\overline{P'Q'} \tag{5.13}$$

which was of course to be expected, but it serves also to check the correctness of the formulae.

Formula (5.10) (together with (5.8) or (5.11)) suggests a projective invariant description of a line segment with regular subdivision points, the points P and Q, the number of subdivisions n plus one additional subdivision point, for example A_1. The formulae dictate that if these and only these points are transformed by the *full* transformation, using formula (5.10) and the corresponding (5.8) or (5.11), the image of all other subdivision points can be generated, and this is exactly what is required ([Herm89]).

Although this approach is basically correct and this is what will be done in general, one has to be a little bit cautious. The arguments presented above were based on the assumption that A_i is an element of the line segment PQ. Indeed, formula (5.3) was transformed into (5.4) by using a substitution of the value $\overline{A_iQ}$. It is easy to see, however, that if A_i happens to be on either side of P and Q (that is, outside the line segment PQ), this substitution is still valid, and that the final formula (5.6) is still usable. This also means that the value of i in the formulae (that is the index denoting the next point) is not necessarily confined to the range of $1 \leq i \leq n-1$; it can, theoretically, be extended beyond n. Furthermore, negative i values are also usable; when evaluating the value of (PQA_i) (see (5.7)) a negative i will automatically give a negative value for (PQA_i), which is exactly what is necessary according to the definitions of the division ratio. In other words, formula (5.6) (with the help of (5.8) and/or (5.11)) describes not only the subdivision points of the line segment PQ but the extensions of these subdivision points in both directions of the line.

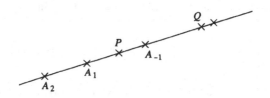

Figure 5.3.

Also, if for example the value of $\overline{PA_k}$ is negative (that is if A_k is for example on the opposite side of P vis–a–vis the point Q), the formula is still correct (see figure 5.3).

However, attention must be paid to the fact, that (5.6) might lead to singularities; that is, ideal points also belonging to the line PvQ should be dealt with. The appearance of this singularity is quite normal and it means that the situation when the line segment PQ becomes external should also be handled properly. As stressed in the introduction, no W–clip is supposed at this point (although the very fact that one of the points P, Q or A_k may be ideal must be and can be considered as known; this can be seen easily when performing the projective division).

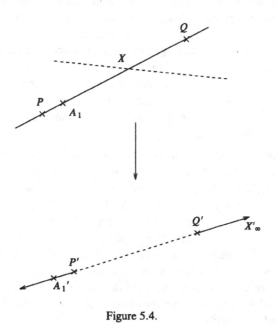

Figure 5.4.

To see how this problem fits into the formulae above, the case when the whole PvQ line is ideal might be disregarded for the time being. This means that at most one point among P', Q' and A_i' is ideal. Also, the case when one of the points P' or Q' is ideal will be postponed until later. The fact that the image of the PQ line segment becomes "external" means in practice that the intersection of the line segment PQ and the vanishing plane (or line) of the transformation intersects in an interior point X of PQ. The transformation maps the point X onto an ideal one; consequently the image of the line segment PQ will be the set–theoretic complement of the line segment $P'Q'$ (see figure 5.4). It is also clear that the position of A_1' (or, in general, A_k') will be outside the line segment $P'Q'$ (this is the result of the invariance of the segment cutting property). This is also true conversely: the very fact that A_k' is not on the line segment $P'Q'$ means that a W–wraparound has been in effect.

The initial position of A_k' can be found easily; there exists a value τ for which

$$A_k' = \tau P' + (1 - \tau)Q' \tag{5.14}$$

and:

$$\tau < 0 \quad \leftrightarrow \quad A_k' \text{ is on the half line of } P'$$
$$0 < \tau < 1 \quad \leftrightarrow \quad A_k' \text{ is on line segment } P'Q' \qquad (5.15)$$
$$1 < \tau \quad \leftrightarrow \quad A_k' \text{ is on the half line of } Q'$$

In the situation shown in figure 5.4 the application of formula (5.6) will generate automatically the images of the subdivision point which will, in this case, go *away* from P'; at a certain time a singularity occurs and the points will afterwards appear on the half line generated by Q'. This is clear from the formulae: the value of $\overline{P'A_i'}$ will be negative for a while then it will "jump" onto a (usually very large) positive number. Using negative indices one may also generate the "internal" points, as in figure 5.3. The exact interval $A_l'A_{l+1}'$ where the effective wraparound occurs can also be found as a "by–product" of the formulae in use (this by–product will, however, be very important later).

The fact that formula (5.6) has a singularity means also that the values generated by the formula (which describe the directed distances from P) can be very large and can therefore exceed the largest floating point number which can be managed by the computing environment. This is however not a real practical problem; all graphics output generation has to be clipped ultimately by the workstation clipping cube and therefore such a large number will be disregarded anyway. The two steps can also be combined: once the generated A_i' has exceeded the workstation clipping cube limits, this maximum value can just be stored. Furthermore, by inspecting the formula (5.6) and knowing the limits of the workstation clipping cube, the values of the indices where the corresponding points will be out of the limits could be found very easily.

Precisely speaking what has been described before corresponds to what is shown in figure 5.4. The point A_k' can also be on the other side, that is on the half–line generated by Q'. This case corresponds to the fact that the point X, which is on the line segment PQ is also on the line segment PA_k. This might be the source of real problems, however: if this case occurs, the value of $\overline{P'A_k'}$ will be very large and this might lead to serious calculation errors when using (5.6)! However, the situation is absolutely symmetric; one could exchange the role of P with that of Q, the role of Q with that of P and A_k with A_{n-k} to avoid problems. This is also what has to be done if A_k' happens to be ideal.

One more case should be treated separately, namely when either P' or Q' happens to be an ideal point. In this case, the original considerations leading to formula (5.6) are not valid any more; a separate formula should be sought.

110

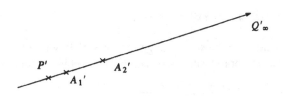

Figure 5.5.

Returning to the original arguments leading to (5.6), the value of (PQA_kA_i) might still be considered as known (and is denoted again by $\alpha_{k,i}$). In this case, however, Q is an ideal point. Using the permutation relations of the double ratio as well as the original definition for ideal points (see chapter 2) the following holds:

$$\alpha_{k,i} = (PQA_kA_i) = (A_kA_iPQ) = -(A_kA_iP) = -\frac{\overline{A_kP}}{\overline{PA_i}} = \frac{\overline{PA_k}}{\overline{PA_i}} \qquad (5.16)$$

that is, by using the formula derived above for $\alpha_{k,i}$:

$$\overline{P'A_i'} = \frac{i(n-k)}{k(n-i)}\overline{P'A_k'} \qquad (5.17)$$

Clearly, formula (5.17) is the alternative for (5.10).

These results can now be summarised as follows. A line segment PQ is given together with an integer n (that is with the subdivision points $A_1,\dots,A_{n-1}$). Additionally, another integer k is also given, which should be $k<n/2$ (preferably, k is equal to 1). If an arbitrary projective transformation is given, the Euclidean image of the subdivision points can be generated using the following procedure:

i) The points P,Q,A_k,A_{n-k} are transformed, producing the (possibly ideal) points P',Q',A_k',A_{n-k}'. If both P and Q are ideal, the whole line will be ideal and therefore the image of the subdivision points cannot be represented on the Euclidean plane.

ii) If Q' is ideal, then apply the following formula:

$$\overline{P'A_i'} = \frac{i(n-k)}{k(n-i)}\overline{P'A_k'} \qquad (5.18)$$

to generate the A_i'-s.

If P' is ideal, the same formula should be used by exchanging the roles of P' with Q' and A_k' with A_{n-k}'.

iii)   If neither Q' nor P' are ideal, find a value τ for which:

$$A_k' = \tau P' + (1 - \tau)Q' \quad (\tau < 1)$$

If such a τ does not exist, exchange the roles of P' with Q' and A_k' with A_{n-k}' (in this case an appropriate τ will exist). Then apply the following formula:

$$\overline{P'A_i'} = \frac{(P'Q'A_k')}{\alpha_{k,i} + (P'Q'A_k')} \overline{P'Q'} \tag{5.19}$$

where the value of $\alpha_{k,i}$ is:

$$\alpha_{k,i} = \frac{k(n-i)}{i(n-k)}$$

Precisely speaking, the formulae above will give the directed distances between the points P' and A_i'. To generate the points themselves, these distance values should be combined with a unit vector parallel to the line $P'Q'$. Both formulae above involve a multiplicative factor and the distance of two points; the equivalent vector equations become therefore fairly straightforward. Thus, if case *ii)* described above is valid then:

$$\vec{A_i'} = \vec{P'} + \frac{i(n-k)}{k(n-i)}(\vec{A_k'} - \vec{P'}) \tag{5.20}$$

is the vectorial counterpart of (5.18), whereas

$$\vec{A_i'} = \vec{P'} + \frac{(P'Q'A_k')}{\alpha_{k,i} + (P'Q'A_k')}(\vec{Q'} - \vec{P'}) \tag{5.21}$$

corresponds to (5.19).

Some remarks are of interest. Clearly, whether both A_k and A_{n-k} are to be transformed or not, depends on the problem of W–wraparound. If no wraparound occurs (which can be found out by using a kind of "cheap" W–clip by just comparing the w coordinates in 4D), one of the transformations is superfluous. An implementation might want to make use of that. Also, the simple fact that

$$(P'Q'A_k') = \frac{\tau}{1-\tau} \tag{5.22}$$

also holds in case of *iii)*, might save some calculations.

Finally, both in steps *ii)* and *iii)*, the roles of the points may be exchanged. When actually implementing the method, care should be taken to use a correct indexing of the results.

It is also clear that a step towards a more formal specification of an extended 2D system might be to define a line segment by the points P',Q',A_k',A_{n-k}', since these points provide enough information (together with the values of n and k, of course) to generate all other points.

112

5.2.2. Cell Array

A Cell Array is, in its geometric structure, a kind of a two dimensional regular sub-division of a parallelogram; it is therefore natural to apply the method described previously to render it after the projective transformation.

The aim is to generate on the projective plane all the grid points, or at least those of them which are affine. Figure 5.6 shows the original grid; figure 5.7 shows one possible image of 5.6, which is also the simplest one.

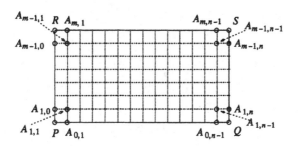

Figure 5.6.

The points which are to be transformed are shown in figure 5.6. Using the method described in the previous chapter, first the images for the subdivision points on the four edges on the parallelogram can be calculated. As a next step, the first and the last column of the grid (that is the points $A_{1,1}', A_{2,1}', \ldots, A_{m-1,1}'$ as well as the points $A_{1,n-1}', A_{2,n-1}', \ldots, A_{m-1,n-1}'$) and, finally, each row can be generated one after the other.

It must be stressed that, as stated before, the factors denoted by $\alpha_{k,i}$ are all the same for each "horizontal" as well as for each "vertical" line respectively. In other words, once these factors have been calculated (in the first step), they can be reused to shorten the calculations. Also, the choice between generating first the two columns and then the rows rather than the other way round is arbitrary; theoretically at least, there is no difference between the two choices as far as the generation of the grid points is concerned.

Figure 5.7 shows the simplest (and most probable) case, in which no wraparound happens on the parallelogram during the projective transformation. In this case not all the points listed before are to be transformed; the generation method for the subdivision points can be used without the points $A_{0,n-1}', A_{m,n-1}', A_{m-1,0}'$ etc. To check whether a wraparound occurs can easily be done by comparing the w coordinates of the vertices before projective division.

If a wraparound does occur, all points are necessary; however, using the method for the generation of subdivision points in general, all possible images of the

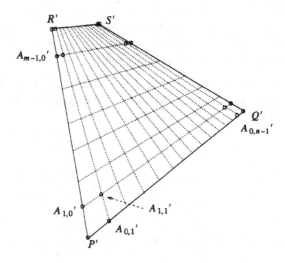

Figure 5.7.

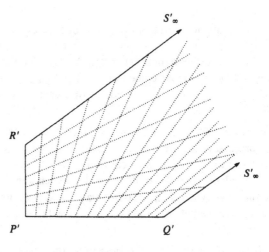

Figure 5.8.

regular grid are automatically generated. Figure 5.8 shows a case when S' becomes an ideal point; in figure 5.9 both the line segments $P'Q'$ and $R'S'$ produce a W-wraparound (in other words the vanishing plane intersects the line segments PQ and RS) and, finally, the line $Q \vee S$ is fully on the vanishing plane on figure 5.10.

Precisely speaking, the generation described above produces the set of grid points only. In the simple case where no wraparound is produced (figure 5.7) or even if none of the interior points becomes ideal (eg figures 5.8 or 5.10) the generated grid points effectively determine the image of the corresponding cells. In other words, after having determined the points:

$$A_{p,q}', A_{p,q+1}', A_{p+1,q}', A_{p+1,q+1}'$$

these points may be considered as being the four vertices of a cell; by assigning a given colour to it, it can be treated by the rest of the output pipeline as a solidly filled quadrilateral.

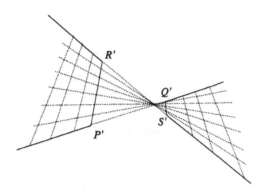

Figure 5.9.

In the case, however, where the vanishing plane intersects the interior of the parallelogram *PQRS*, this is no longer true. It may happen that (using the previous notations) the line segment $A_{p,q}'A_{p,q+1}'$ produces a W–wraparound, that is the image of the corresponding cell is *not* the quadrilateral. Examples for this situation can be seen in figure 5.9.

As described in the previous chapter, it is very easy to detect for a given line segment which of the subdivision intervals would produce a singularity. Basically, this happens when the sign of the directed distance $\overline{P'A_i'}$ differs from the sign of $\overline{P'A_{i+1}'}$; it is therefore possible to find out which original cells would produce a W–wraparound.

Two cases should be differentiated. The first one is when the vanishing plane of the transformation intersects two parallel edges of the cell parallelogram (see figure 5.11), say the edges *PQ* and *RS*. In this case the grid points on the image should be generated in rows, that is the grid lines which originate from the line segments *parallel* to *PQ* should be calculated (remember that choosing this "row" or the alternative "column" approach leads to the same results). As each sub–parallelogram which intersects the vanishing plane will also intersect it on at least one row (that is a "horizontal" edge), all quadrilaterals leading to a

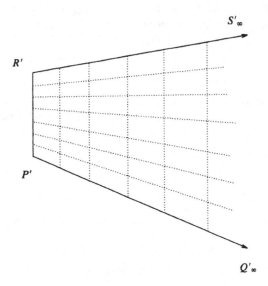

Figure 5.10.

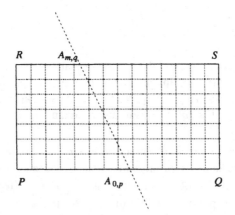

Figure 5.11.

W–wraparound will be detected this way. Of course, if the vanishing plane intersects the edges PR and SQ, the same approach should be used by taking columns
instead of rows.

If the vanishing plane intersects two adjacent edges (figure 5.12), the situation
is more awkward; if for example the intersection points are on the edge PQ and PR

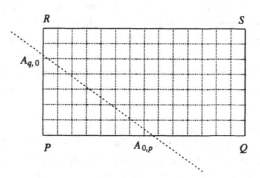

Figure 5.12.

at the index p and q respectively, all grid points $A_{i,j}'$ for which $i \leq p$ and $j \leq q$ should be generated in both column and row direction to find the possibly singular quadrilaterals. Indeed, neither of the two directions may alone detect all cases.

To find out which of these two cases is actually valid is an easy task; it is a by–product of the first step of the algorithm, when all grid points on the edges are generated.

The final question is: what should happen with a singular quadrilateral? To put it very pragmatically, in most cases such a quadrilateral could just be disregarded. The fact that an ideal point is involved means that there is a point on the corresponding edge for which the directed distance involved in all formulae becomes infinite. In other words, it will be so "far away" that with very high probability this quadrilateral will be outside the workstation clipping cube anyway. This can be checked very easily: if all four vertices are outside the workstation clipping cube, the image of the corresponding cell (whatever it looks like) will be outside as well.

In case a more precise solution is sought, a standard recursive procedure will produce the visible subparts. By putting $n \rightarrow 2n$, $m \rightarrow 2m$, it is possible to generate an imaginary grid which would be twice as dense as the original one (on the original image prior to the transformation); consequently, the quadrilateral can be divided into four sub–areas repeatedly. However, this recursive part of the algorithm should be present as a kind of a security measure only; it will hardly be used in practice. It has to be stressed as well that, in fact, in most practical cases the singularity problem does not occur at all.

When comparing the cell array generation in 4D with the one described here, one can count the number of floating point operations needed to generate the next point in the grid. In this case, the number of operations is slightly higher for the projective algorithm than for the 4D version described in chapter 2 (8 versus 7). In other words, this method is not faster than the 4D approach.

The projective algorithm can also be compared to a purely *planar* projective transformation of a cell array; in this case, the projective algorithm described here is clearly superior in speed. An experimental implementation in C++ has shown a speed improvement in the range of 20%–25% (depending on the size of the array); an improvement which is primarily due to the fact that the number of floating point operations has been reduced in case of the projective version. Taking into account that the cell arrays can be quite large, this 20% gain can have a significant practical importance.

The reason to look for a projective algorithm in the case of cell arrays was however not exclusively speed; clearly, it also provides a way to define an extended two dimensional graphics system in a compact way. It might also lead to some other interesting applications; these will be presented in the chapter dealing with possible directions of future research.

5.2.3. Pattern Filling

Pattern filling is very similar to drawing a cell array. A regular grid is defined, with the difference that the basic pattern is duplicated in all directions to form a (conceptually) infinite grid.

A regular grid can be reproduced after the projective transformation: as stressed in relation to the mapping of regular subdivision points, the indices for the subdivision points A_i'-s may extend onto the whole set of natural numbers. Based on this fact one way to perform the pattern filling is as follows.

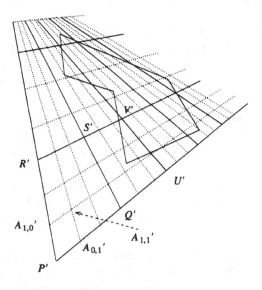

Figure 5.13.

Figure 5.13 shows a regular pattern, generated by a relatively small (that is 3×3) pattern array. The main target is to find finite index intervals both in the PQ and the PS directions so that the whole polygon should be contained within the (usually larger) regular grid determined by these intervals. In a more precise form this means that four integer values i_0, i_1, j_0, j_1 should be found so that the polygon itself is contained by the sub–grid determined by the subdivision points

$$A_{j_0,i_0}, A_{j_0,i_1}, A_{j_1,i_0}, A_{j_1,i_1}.$$

In figure 5.13 the choice

$$i_0 = 4$$
$$i_1 = 14$$
$$j_0 = 0$$
$$j_1 = 7$$

will do. If such intervals are found, the rest becomes quite simple: all quadrilaterals determined by the intervals should be generated one after the other; the original polygon should be clipped against these quadrilaterals[†], and, finally, all clipped quadrilaterals are displayed using the appropriate colour determined by the pattern specification.

To find such intervals does not seem very complicated either. If the half space determined by $P'vR'$ contains the polygon, then one can proceed in this direction as long as this condition is still true, to find the value of i_0; once this has been done, further steps can be taken in the same direction to find the value of i_1. If not, the steps have to be done "backwards" to find i_0.

If this approach were feasible, the result would be quite appealing, even when compared to the 4D algorithm described in chapter 2. Although the clipping step itself is more complicated than the one described there (a general convex clipping area should be used instead of a regular rectangle), a number of processing steps can be avoided. In fact, the polygon can undergo all necessary clips needed by the whole output pipeline (modelling clip, workstation clip) before being filled by a pattern whereas in the 4D case each individually clipped quadrilateral still has to be projected back onto the $w=1$ plane and clipped by at least the workstation clip (even if the modelling clip is performed in 4D). As stressed before, clipping is a computationally expensive operation and therefore avoiding a clipping step can be very advantageous.

In some cases the projective version of pattern filling may have even more advantages; one example is the Hidden Line/Hidden Surface calculation. Although

---

[†]Not the other way round! Theoretically the quadrilaterals could be clipped against the polygon as well, but whereas the quadrilaterals are convex polygons, the original polygon is not necessarily one. On the other hand, clipping against a convex polygon is faster than against concave ones.

the so-called Z-buffer algorithm is one of the most widespread approaches to solve this problem nowadays, it is by no means the only and necessarily the best one. Indeed, the Z-buffer method presupposes that the image is generated on some kind of a pixel-based output device which is not always the case. A trivial example would be the use of pen plotters, but even in the case of displays there are attempts to produce pictures directly from a higher-level description and to by-pass the use of pixel memory which has proven to be a bottleneck in a number of cases (see eg [Ghar85] or [Hage87]). If this is the situation, the Hidden Line/Hidden Surface problem must be solved with the help of some software algorithm (Newell-Newell-Sancha or the like) by producing the visible part of the 2D projection for each polygon. The fact that this step can be done *before* using any pattern filling at all if the projective algorithm presented here is used may be of importance.

However, care should be taken when applying the method described above; as usual, the existence of singularities might lead to problems and should be handled properly.

First of all, it must be supposed that the polygon itself does not contain singularities any more. In other words, either the W-clip or the UW-clip should be considered as already done (in any other case, it would be too complicated to find out the singular points within the polygon). Furthermore the fact that the workstation clip can also be performed as the first step means that the polygon itself may be considered as bounded[‡]. This is clearly necessary to be able to close the polygon into a finite part of the grid.

The fact that the original pattern array (like the cell array in the previous chapter) might be intersected by the vanishing plane of the transformation is the source of a number of complications, mainly for the indexing of the subdivision points. Instead of transferring these complications to the pattern filling as well, it seems much simpler to suppose that the basic pattern array does *not* intersect the vanishing plane. Indeed, it has no effect on the final picture if the defining array is translated into any of the two main directions; following the notations of figure 5.13, the point Q' might play the role of P', U' might be used as Q', V' as S' and finally S' as R'. This translation of the pattern definition can be done very easily in 4D; if the original parallelogram happens to intersect the plane $w=0$, it can be translated in a given direction to avoid the intersection. This will not change the generated grid but makes the indexing of the subdivision points simpler.

However, unclear situations may still arise, and to avoid them a more exact description of the grid is necessary.

Figure 5.14 shows the (strongly distorted) image of a grid generated by a 3×4 pattern array, denoted by P', Q', R' and S'. The basic pattern is duplicated in all directions. What happens is that a regular grid defined by the pattern is generated by

---

[‡]Even if the original polygon is bounded, its projective image can still be unbounded, so this remark is necessary.

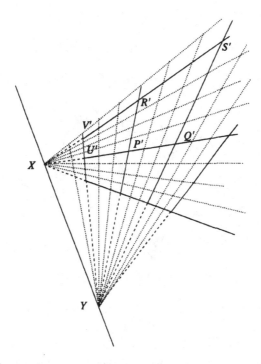

Figure 5.14.

all possible intersection points on the projective plane by two well–defined bunches of lines. These bunches of lines have the common intersection points X and Y respectively and the lines belonging to the respective bunches are governed by the rule that their respective double ratio should be the values of $\alpha_{j,i}$ of formula (5.8). For a regular grid the points X and Y are the ideal points determined by the lines $P \vee Q$ and $P \vee R$ respectively; such a bunch of lines is then transformed by the projective transformation into a more general configuration where the points X and Y are not necessarily ideal points any more[t]. Clearly, the line $X \vee Y$ is the image of the ideal line.

As seen on the figure, all pattern quadrilaterals "converge" in some sense toward the points X and Y. By performing a more detailed analysis of the values of α_i as well as of the value of $\overline{PA_i}$ in formula (5.6), this fact can be described precisely, but all these details are not really of importance here[‡]. It is sufficient to say that, based on the value of $(P'Q'A_i')$ either the subdivision points starting from P' in

---

[t] Such configurations are also called "Möbius Nets" in projective geometry.

[‡] Basically, the values of $\lim_{i \to +\infty} \overline{PA_i}$, $\lim_{i \to -\infty} \overline{PA_i}$ as well as the change of sign of $\overline{PA_i}$ should be described in more detail.

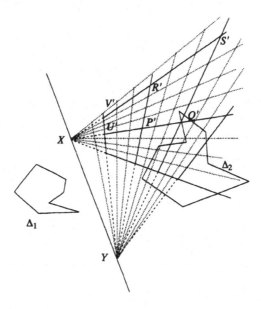

Figure 5.15.

direction U' (figure 5.15), or the ones in direction Q' will converge towards X without leaving the half–plane defined by $X \vee Y$ and P' itself. Points in the other direction will, after a certain number of steps, "swap" to the side of $X \vee Y$ opposite to P'. This also means, however, that if the polygon happens to be on the *other* side of $X \vee Y$ (like Δ_1 in figure 5.15), the stepwise approximation described at the beginning may never come to an end; indeed, in figure 5.15 there will be no index i_0 which would give a lower bound for the polygon (the corresponding lines will never "cross" $X \vee Y$). There will be no problem, however, if the polygon is on the same side as P', Q', S' and R' (which is the case of $\Delta_2{}^\dagger$). Consequently, it has to be secured that when performing the pattern filling itself, the polygon to be filled and the defining pattern quadrilateral should be on the same side of $X \vee Y$.

It is not particularly complicated, however, to achieve this. When performing the W–clip, all primitives will be either on the $w > 0$ or the $w < 0$ side. It can be proven easily (see also figure 3.6) that one of these half–hyperspaces will be mapped onto one half–plane of $X \vee Y$ and the other half–hyperspace to the other one. The only step to be done is to translate, if necessary, the defining pattern before projective division so that the whole pattern lies in the same half–hyperspace of $w = 0$ as the polygon itself (remember that in most of the practical cases, this will be true

---

†Clearly, neither the polygon nor the defining pattern quadrilateral will intersect $X \vee Y$; indeed, the original definition for these point sets does not include any ideal points.

automatically). As this translation has to be done anyway to avoid singularities (or at least the necessity of it should be checked) this does not produce any additional computational complications.

There is still another source of problems, but this is much more theoretical than practical. As said before, one direction of the subdivision points will "converge" towards the point X or Y, while the other side will "swap", after a certain number of steps, to the other side of $X \vee Y$. In figure 5.15, starting from P' and going towards positive indices will produce this latter effect. This also means that in a very unlucky case there will be no upper bound i_1 generated for the polygon, even if it is, like Δ_2, on the right side of $X \vee Y$. This will correspond to the already described situation when W-wraparound is produced by the next subdivision point generation.

Just as in the case of cell array, however, this problem is more theoretical than practical and can be detected with a small number of steps (in contrast to the previous problem which might have lead to an infinite loop). It is theoretical, because the problem will occur usually on very distant locations vis-a-vis the workstation clipping cube. Just as with the cell array, an emergency measure can be to continue with a more dense series of subdivision points to find an appropriate i_1; however, this will hardly be necessary in practice.

With all these precautions the method described above to perform pattern filling in a projective environment can be applied. Clearly, in a more formal specification, a pattern should be described more or less like a cell array (with all four vertices plus the internal "corners").

5.2.4. STROKE Characters

As previously stated, STROKE precision characters are defined on a regular grid as well. In fact, the final image of a whole string is a superposition of two regular grids: a relatively coarse one which corresponds to the character boxes themselves, and a much finer one which constitutes the grid of the character description (see figure 5.16).

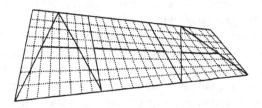

Figure 5.16.

The reproduction of this grid may follow exactly the same method as described earlier. In practice, a two-step process might be advantageous: by generating the coarse grid first (that is the character boxes) a kind of a pre–clip can be performed to see whether the character at hand is visible at all. Remember that a good character description might consist of a relatively high number of points, so it is worthwhile saving some clipping if possible.

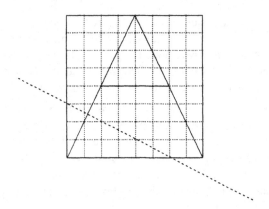

Figure 5.17.

There are, however, two remarks which have to be made about character generation. The first (and usual) problem is how to handle singularities. Unfortunately, for STROKE characters it is not possible to do this properly. The reason is that in contrast to the previous problems, the (transformed) grid is not used "cell–based", that is instead of handling the individually generated grid quadrilaterals, the usual description of characters use the grid as a kind of special coordinate system, where a line should be drawn between grid points. Figure 5.17 shows the problem arising. The vanishing line of the transformation will cross the character box. The projective generation of grid point will be able to detect that, say, a quadrilateral determined by the grid points $(1,2)$, $(2,2)$, $(2,3)$ and $(1,3)$ will lead to singularity problems after the transformation; this feature of the projective generation has already been used before. However, the generation will *not* detect that the line connecting the point $(0,0)$ and $(4,8)$ and being part of the description of the letter "A" will effectively cross this quadrilateral. In other words, for each individual line such a test should be performed. This is certainly possible, but not really necessary. Remember that a character box intersecting the vanishing plane/line will be very distant anyway; it is perfectly acceptable to just simply rule out these characters from the string (this is not exactly what the STROKE character precision of the GKS–3D/PHIGS specification demands, but in the overwhelming majority of all practical cases it will be enough).

The other, and not really major, problem is related to the more formal specification of an extended 2D system. In all previous cases the points which are to be transformed, that is the points which would serve as a basis for a formal

124

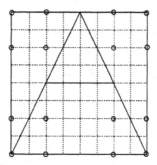

Figure 5.18.

specification are the ones shown in figures 5.6 and 5.7; they are essentially the corner points. However, the exact resolution of the character description depends on the character font in use; furthermore, the user of such systems is not entitled to know the value of this resolution (which was 8 for all of the figures, like 5.18). In other words, the internal points defined in this case should correspond to some geometrical feature: for example the ½ and ¼ division points of the character box should be used (see figure 5.18) and the system itself may then find out that these points correspond to the indices, say, 2 and 6. This also means that the more general form for $\alpha_{k,j}$ should be used in the corresponding formulae.

5.3. Conics

Handling of conics in a projective invariant manner is quite different from what has been described for cell array, pattern filling etc. Whereas in all these cases a general method was used with some special arrangements for handling singularities, in the case of conics the singularities themselves determine the global geometric (and affine) behaviour of the curve, and this feature will be exploited.

It has been shown in chapter 3 that for each class of conics one can assign a set of *characteristic points*; these points have the property of generating parametric formulae which can be used to approximate the curve. Furthermore, these formulae are such that they can also describe arcs instead of complete curves, provided that the starting point, the end point and an interior point of the arc are also given. Taking these facts into account, it is a natural idea to try to find a description for the conics such that:

- the description should be easily transformed by a projective transformation in an invariant manner;

- it should be possible to construct this description out of the characteristic points;

- it should be possible to reconstruct the characteristic set of points out of this description.

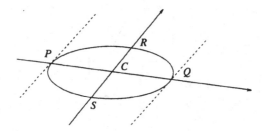

Figure 5.19.

It is also worthwhile recalling here what the characteristic points are.

- For an *ellipse* (figure 5.19), the set consists of the centre (*C*) and the endpoints of two conjugate radii (*R* and *Q*).

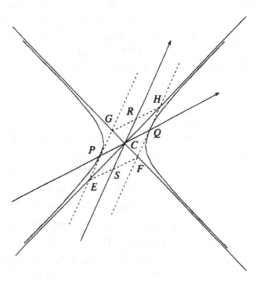

Figure 5.20.

- For a *hyperbola* (figure 5.20), the set consists of the centre again (*C*), the two intersection points of a chord (*P* and *Q*) and two points (*R* and *S*) of the line which is conjugate to the chord and contains the centre. The precise way the points *R* and *S* are specified has been detailed in chapter 3 and will be

126

re–explained later in this chapter as well.

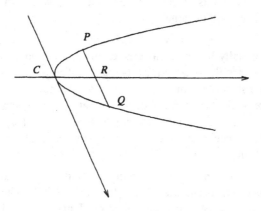

Figure 5.21.

- For a *parabola* (figure 5.21), the set consists of the intersection point of a diameter (*C*), the two intersection points of the curve and a chord conjugate to the diameter chosen before (*P* and *Q*) and, finally, the intersection point of the diameter and this latter chord (*R*).

How all these points are constructed has already been presented; what will be shown in the present chapter is that from the matrix and some additional data these points can be *calculated*. A number of basic calculation steps in relation to conics have been well described previously (and were presented in chapters 2.9.2.1 and 2.9.2.2). For example, if the homogeneous coordinates of two points are known, it is possible to calculate the homogeneous coordinates of their generated line as well those of its pole. Also the polar of a point can easily be specified and this polar will be just the tangent of the curve if the point is on the curve. All these calculations will be used in what follows, but their details will not be repeated here.

It has also been mentioned in chapter 2 that the *matrix* of a conic is, in a sense, an invariant description of the curve, and, if the matrix of the conic is denoted by A and the projective transformation by T, the relationship

$$T(A) = (T^{-1})^T A (T^{-1}) \qquad (5.23)$$

gives the matrix of the image of the conic under the effect of T. In other words, the matrix of the curve is *projective invariant*. It is therefore a natural idea to use the matrix as a starting point for the required description.

It is also fairly easy to construct the matrix of the curve, once its classification (that is whether it is an ellipse, a hyperbola or a parabola) and its characteristic points are known. This was true both for 2D as well as for 3D (in the latter case a generalised cylinder was used to represent the matrix). Unfortunately, the converse is not true. It is not easy to generate characteristic points out of the matrix; in fact, it is not even easy to find at least one affine point at all! It seems necessary to transform at least some points of the curve as well, to have a starting point for the generation. This is what will be done.

A conic will be represented by its matrix and *three* of its points. Clearly, if a curve is defined originally by its characteristic points, three points of the curve can be generated without problems as well as its matrix. The projective transformation has to be used to transform the points themselves and formula (5.23) may be used to generate the matrix of the transformed curve. What has to be shown is that if the matrix A of the curve and three arbitrary points of the curve (say, X, Y and Z) are given, it is possible to reconstruct the characteristic points. The necessary steps to achieve that are as follows.

i)   If the problem is given in 3D, the points X, Y and Z can be used to determine the homogeneous coordinates of the plane $\Pi = X \vee Y \vee Z$. This plane contains the curve; also, at least two ideal points of Π may be considered as known (see chapter 2.9.2.2.)

ii)   The conic has to be classified. This means that the intersection of the curve and the ideal line (of the plane) had to be found. The method of doing this was presented as a proof of theorem 2.26 for 2D and generalised for 3D in chapter 2.9.2.2. As a by–product, not only is the classification achieved but the eventual ideal points of the curve are also calculated.

iii)   The centre of the curve has to be found. If the previous step has led to the fact that the curve is a parabola, then this is automatically at hand: the (only) ideal point is also the centre of the curve. If this is not the case, the pole of the ideal line has to be calculated (remember that the center is the pole of the ideal line). This can be done by choosing two ideal points (I_1 and I_2) and

$$(AI_1) \wedge (AI_2) \tag{5.24}$$

will generate the pole in 2D (AI_1 and AI_2 describe the polars of the points I_1 and I_2 respectively). In case of 3D, the intersection with the plane Π has also to be added to (5.24).

iv)   In case the curve is an *ellipse*, one of the known points of the curve may be assigned as Q. The pole of the line $C \vee Q$ can be calculated using (5.24) again (in fact, this pole will be an ideal point). This ideal point will determine the direction of the line conjugate to $C \vee Q$; its intersection points with the curve can be determined using (as in *ii)*), the proof of the theorem 2.26. One of the intersection points will be R.

v)   For a *hyperbola*, the situation is slightly more complicated. Taking one of the affine points of the hyperbola and denoting it by P (at least one of the points X, Y or Z should be affine!) the other intersection point of the corresponding diameter may be calculated. For each of the two asymptotes two points of it are already known (the centre and the two corresponding ideal points; indeed, the ideal points of the asymptotes are the intersections of the curve and the ideal line, intersections which have been calculated in *ii*)). With the help of these data the points E, F, G and H (intersection points of the asymptotes and the tangents at the diameter endpoints) may also be computed. The method for doing that was presented at the end of section 2.9.2.1, (see also figure 2.20). Furthermore, the diameter, which is conjugate to the diameter PQ can also be determined: the pole of PvQ (determined by the equivalent of formula (5.24)) gives a second point of it (the first being the centre). Finally, the intersection points S and R may be calculated.

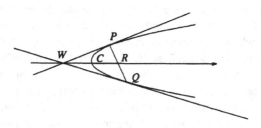

Figure 5.22.

vi)   For a *parabola* two affine points of the curve are already known (out of X, Y or Z), which will play the role of P and Q. The pole of the line PvQ (denoted by W on figure 5.22) may be connected to the (known) ideal point of the curve; this will be a diameter. Based on the remark made in connection with figure 2.19, the intersection of this diameter and the line segment PQ will be the middle point of PQ. Additionally, the (other) intersection point of the diameter with the curve may also be calculated; this will be C, the last missing characteristic point.

It has been shown that using the matrix and three points of a conic the characteristic points can be reconstructed; this also means that this representation is suitable for the purposes of an extended graphics system.

6. Conclusions

The previous chapters have proven that using the tools and the descriptive power of projective geometry, a number of problems can be solved in a more elegant and efficient way and, furthermore, some of the problems arising in 3D graphics systems have been solved for the first time by using these mathematical tools in a consequent manner.

The W–clip (see chapter 3), published for the first time in [Herm87], was the first fully general solution for the W–wraparound problem. Although alternative approaches had been used previously, all had been developed for special forms of projective mappings only (primarily the mappings realising the synthesised camera model). In spite of its simplicity, the W–clip can handle the problem for all projective mappings, regardless of their possible special format. This generality is of particular importance when implementing such 3D packages as GKS–3D or PHIGS. Furthermore, the optimisation methods, developed first in [Herm87] and extended in the present thesis, result in reducing the amount of calculations to their possible minimum; by just inspecting the matrix of the transformation, the number of clipping steps can be reduced. The important point is that this optimisation will automatically lead to reduction of calculations if the transformation happens to realise the synthesised camera model.

The mathematical background used for the development of the W–clip has also lead to additional improvements in the implementation of a 3D graphics pipeline. By moving some calculations into the four dimensional Eucledian space instead of performing them prior to the projective transformation, a number of compute–intensive graphics algorithms can be performed much more efficiently than by other means. Examples presented in chapter 3 were Cell Array, Patterned Area Filling, Stroke Precision characters and, with the help of some well–chosen affine invariant formulations, all conic curves.

The modelling clip problem was, for some time, one of the unsolved algorithmic problems of 3D computer graphics. The use of the four–dimensional geometry provided the first viable solution for this problem, in spite of the fact that this approach could be used in limited, albeit very important cases only. This solution, presented first in [Herm88], has also been fully described in chapter 4. The full mathematical analysis of the modelling clip problem with solutions provided also for the most general case, was also described in that chapter. The most important result presented there was the fact that the modelling clip can be performed after the full projective transformation, by performing a clip against two well–chosen convex bodies.

By using very different techniques of projective geometry, new and alternative approaches for the fast implementation of some graphics algorithmic problems were presented in chapter 5. These methods have in common the idea of finding a *projective invariant* formulation for some graphical output primitives. Beyond the possible speed improvement offered by these methods, these formulations have the additional advantage of offering a new platform for the specification of new, general and application–independent graphics systems. In chapter 5, the necessary mathematical tools, based on the notion of double ratio, were developed and the algorithmic methods presented for the case of such primitives as Cell Arrays, Filled Patterned Areas and Stroke Precision characters. Also, projective invariant formulae and generation methods were found and developed for the approximation of conics under the effect of projective transformations, making use again of the projective theory of conics.

7. Directions for Further Research

Beyond the specific algorithms which have an interest and importance of their own, the present study was aimed at communicating a more general message. This might be considered to be an *attitude*: it says that the use of more complicated and more advanced mathematics might have fruitful results when applied to some relatively well–known computer graphics problems and, in some cases (eg modelling clip), such an approach may be the only one which leads to correct solutions. What "more advanced" means is of course relative; results presented here cannot be considered as real and deep novelties for pure mathematicians. If, however, the general level of mathematical knowledge used in computer graphics is considered and this level is judged upon the current curricula at universities as well as on what is reflected by the usual and widespread computer graphics textbooks (see all the references made before) one can definitely have the feeling that this level is not high enough or, to be more exact, a higher level might lead sometimes to improvements in practice. As such, this attitude might be important in a whole range of problems where projective geometry may have an importance, such as the fields of computer vision, image reconstruction, 3D interaction, 3D input tools (eg 3D locators), stereo image generation and stereo vision. This approach is also valid when using other mathematical fields than projective geometry. The relatively recent emergence of the use of quaternions in computer graphics and animation (see eg [Seid90]) or the use of stochastic processes for modelling terrain (see eg [Anjy90]) are also additional examples.

To be more concrete, the description used in the previous chapters for the images of regular grids (that is what are called Möbius nets in mathematics) may be an interesting starting point for further research. Indeed, this description and the generation methods presented there might be used for the description of the distortions on pixel images as well. In other words, the content of an image memory could be transformed by a two dimensional projective transformation to describe a projective effect. An experimental implementation of the method described in the previous chapter has shown that if the projective algorithm is used to transform a pixel array instead of a "brute force" matrix–vector multiplication plus projective division, a speed gain of 20% can be achieved. Taking into account the very high amount of data involved, this speed improvement might have a great importance. Also, in most the cases, such application do not suffer from the complications related to the appearance of singularities in the algorithms, which makes them even simpler to use. However, problems may arise when applying this generation method directly. Indeed, the fact that pixels are at discrete locations would require some delicate consideration. Also, a proper sampling of the colour values (by using for example stochastic methods) is necessary to avoid aliasing.

Why is this interesting or important? In animation systems a commonly employed technique is that of key frame animation. Here, important frames in a sequence are generated directly from a geometric description, whilst intermediate ones are derived by interpolation. In the course of such an interpolation it might be interesting to have a method which would be able to distort a frame (that is a pixel image) in a projective sense directly. Similar kind of distortions appear in texture mapping (see for example [Heck86] or [Heck89]) where the mapping of textures onto the screen may be described by a two dimensional affine or projective mapping of the original texture.

Another example of an application where such pixel distortions might be important is related to systems where traditional computer graphics (which might also be called "image synthesis") and image processing (or "image analysis") are used together. A major issue today is to see if it is possible to create a unified discipline which would encapsulate both image analysis and synthesis; in their recent survey in [Pun90], Pun and Blake have proposed the name "imagery" to cover such an area. In such an environment, realistic images may be constructed by mixing synthesised images (following traditional methods using some geometrical database) and direct pixel images (like photographs for the background). If such images are then to be seen from different viewpoints the original photograph must be distorted again which leads to problems which might be related to the approach described in the previous chapter.

A completely different field for possible further research is as follows. As explained in chapter 2, projective geometry has resulted in a whole series of tools and methods used by draughtsmen to produce technical drawings. These methods include tools for drawing the image of a projected circle, to generate the image of the middle point of a line segment etc. In some countries this field has become a separate discipline within mathematics as some kind of special chapter of projective geometry, called "Descriptive Geometry" in English (or "Darstellunggeometrie" in German) while in other places it is part of a larger discipline concentrating on technical drawing in general (which involves a whole range of additional problems which have nothing to do with projective geometry); [Fäus71] is just one of numerous possible references for such textbooks. These methods are powerful and in most cases relatively easy to use; as an example, most of the figures in this study have been generated using such methods. In some way, the generation of Möbius nets as described in the previous chapter has also been inspired by these approaches and the reconstruction of the characteristic points of conics has also followed similar mental paths.

It is therefore a natural idea to see whether these methods could be used for computer graphics as well. The problem is that in most of the cases these methods are based on the ability of the draughtsman to generate the intersection of two lines with the help of rulers. Whereas this is easy to do for a human, the necessary calculations are not the simplest ones for computers. However, one could imagine a hardware/firmware configuration where the intersection calculations are basic primitives realised by some hardware or firmware. The necessary calculations, which involve the evaluation of some formal determinant, are very regular and therefore easily realisable by an appropriate microcode or hardware. In such cases, however, the step of creating the intersection of two lines becomes cheap which might mean that the computer might try to simulate a draughtsman. Whether such an approach would really lead to significant improvements is not clear but this might be worthwhile to investigate.

It is also a natural idea to follow the investigations on conics toward quadratic surfaces. The reason this has not been done up to now is that generation of quadratic surfaces leads to a problem which is not closely related to the quadratic surfaces proper but is of a more general nature. This problem is the question of rendering *shaded* 3D surfaces. The problem occurs in very practical terms in case of a PHIGS PLUS implementation (although not in using quadratic but Rational B–Spline Surfaces). This problem is presented in more detail in the next chapter.

8. An Unsolved Problem: Shaded B–Spline Surfaces

To make the problem of shaded B–Spline Surfaces understandable, it is worthwhile giving a very short introduction to rational and non–rational B–Spline Surfaces. Apart from the problem to be described it will also be interesting to see that the 4D approach which has extensively been used throughout the present study is also used in the theory of rational B–Splines, even if this is not always clearly stated.

Mathematically, *splines* are piecewise polynomial functions; these functions are used in approximation theory to approximate more complicated functions. In computer graphics, B–Splines are used in a similar sense; however, the aim is rather to approximate a set of points either with a curve or with a surface and in such a way that the approximation should be easy to calculate and to manipulate by modelling and graphics systems.

Definition 8.1. A kth degree (non–rational) B–Spline curve $C(t)$ is defined by:

$$C(t) = \sum_{i=1}^{n} B_{i,k}(t)P_i \qquad (a \leq t \leq b) \tag{8.1}$$

where

- the P_i-s are 3D (or 2D) points, called control points.
- a and b are fixed with $0 \leq a < b$.
- the $B_{i,k}(t)$ are scalar-valued spline functions in the variable t, of order k (degree $k-1$). They are called the B–Spline basis functions and they form, in fact, the basis of an appropriately chosen linear space of spline functions. The basis functions are completely defined by the order k and a knot vector $\{t_j\}_{j=1}^{n+k}$ where

$$a = t_1 = t_2 = \cdots = t_k < t_{k+1} \leq t_{k+2} \leq \cdots \tag{8.2}$$

$$\leq t_n < t_{n+1} = \cdots = t_{n+k} = b$$

The basis functions are non–zero on a finite number of adjacent intervals defined by the knot vector and zero otherwise. If there exists some positive real number d such that $t_{l+1}-t_l=d$ for $k \leq l \leq n$, the knot vector is said to be uniform and the corresponding B–Spline curve is also called a uniform B–Spline curve; it is nonuniform otherwise ([Till83], [Pieg87]).

Similarly, one may define uniform or nonuniform B–Spline surfaces in 3D:

Definition 8.2. A $k^{th} \times l^{th}$ degree non–rational B–Spline surface is defined as follows:

$$S(s,t) = \sum_{i=1}^{n} \sum_{j=1}^{m} B_{i,k}(s)B_{j,l}(t)P_{i,j} \qquad (8.3)$$

where

- the $P_{i,j}$-s form an $n \times m$ array of control points.

- $B_{i,k}(s)$ is the ith basis function of order k, defined by the knot vector $\{s_p\}_{p-1}^{n+k}$ and $B_{j,l}(t)$ is the jth basis function of order l, defined by the knot vector $\{t_q\}_{q-1}^{m+l}$.

There is a rich and elegant mathematical theory for B–Splines. The reader may refer to the excellent book of Bartels, Beatty and Barsky ([Bart87]) or to the well known overview of the theory given in [Fari88]. In these references methods are presented to evaluate B–splines, like the recursive Cox–de Boor algorithm or the Oslo algorithm, and the basic properties of these curves and surfaces are also described. The appendix of the PHIGS PLUS specification also contains the formal evaluation formulae for these primitives (see [ISO89a]). Rendering these objects is also based on these evaluation algorithms: by appropriate methods a large number of points is generated and these points are then used for a polynomial approximation of the curve/surface. In this respect, the rendering process resembles the one used for conics.

As far as the main properties of B–Splines curves/surfaces are concerned the most important one related to the present study is the fact that a B–Spline curve or a B–Spline surface is *affine invariant*. In other words, transforming a full curve/surface (that is all its generated points) is equivalent to transforming the control points only and generating the curve/surface afterwards. Just as in the case of conics, this feature is of a great importance as far as fast generation and rendering is concerned. However, B–Splines curves/surfaces are *not* projective invariant; in other words, the problem of finding a projective invariant formulation for these curves and surfaces is very important.

The answer to this problem is the introduction of the so–called rational B–Spline curves or surfaces. For such purposes, instead of the usual control points *weighted* control points are used. When a curve is to be described in 3D (which is the usual case), each control point $P_i=(x_i,y_i,z_i)$ is "weighted" with a weight value w_i to generate a set of 4D control points $P_i^w=(w_ix_i,w_iy_i,w_iz_i,w_i)$ where $w_i>0$ (!). With these 4D control points a B–Spline curve is defined in 4D by:

$$C^w(t) = \sum_{i=1}^{n} B_{i,k}(t)P_i^w \qquad (8.4)$$

just as in case of non-rational B–Splines. As a next step, this curve is projected back onto the $w=1$ plane by using the projective division and thus resulting in:

$$C(t) = \frac{\sum\limits_{i=1}^{n} B_{i,k}(t)w_i P_i}{\sum\limits_{i=1}^{n} B_{i,k}(t)w_i} \qquad (8.5)$$

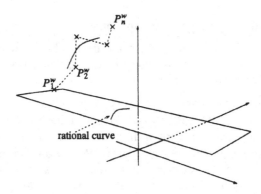

Figure 8.1.

(See also figure 8.1). Similarly, one may define a rational B-spline surface by using (instead of (8.3)):

$$S(s,t) = \frac{\sum\limits_{i=1}^{n}\sum\limits_{j=1}^{m} B_{i,k}(s)B_{j,l}(t)w_{i,j}P_{i,j}}{\sum\limits_{i=1}^{n}\sum\limits_{j=1}^{m} B_{i,k}(s)B_{j,l}(t)w_{i,j}} \qquad (8.6)$$

It is interesting to realise that the use of 4D geometry is not confined to projective problems only. It has to be stressed, however, that rational B–Spline curves or surfaces are *not* defined in projective or even homogeneous terms. Indeed, the requirement that w_i should be positive is clearly in contradiction with the definition of homogeneous vectors[†]; furthermore, the exact choice of the value of w_i has an influence on the shape of the curve and this can be and should be used by designers to adapt the curve to their needs (this is the reason why the value of w_i is called weight; see also [Pieg87]).

The advantages of rational B–Splines over non–rational ones are numerous.

[†]The use of the symbols w_i for weights is the accepted use in the theory of NURB-s, although their use might be misleading in this context, as it suggests that they correspond exactly to the fourth element of a homogeneous vector of projective geometry.

First of all, they allow the generation of surfaces and curves (eg circle, torus, surfaces of revolution) which are not exactly describable by non–rational curves or surfaces. Nice examples can be found in both [Till83] and [Pieg87]. As already mentioned, they give an additional freedom for the control of the final shape via the value of the weight which may be very useful for designers. Furthermore, they seem to offer a way to deal with projective invariance.

It is the case that the rational B–Spline formulation gives a more–or–less projective invariant description of splines. Formula (8.4) defines in fact a *non–rational* B–Spline curve in 4D. To use the linear part of the projective transformation it is therefore enough to transform the control points again and the projective division (that is the step from formula (8.4) to (8.5)) can be done afterwards. In other words, handling a B–Spline via rational curves neatly fits into the frame described in the chapter dealing with 4D methods: the approximation of the final curve (or surface) can be done after the linear part of the transformation and before the projective division.

What, then, is the problem? Shading. Indeed, in modern systems as well as in the more modern standard proposals like PHIGS PLUS, all surfaces should also be rendered (on demand) using some kind of shading. As long as only a flat rendering of a surface is required the above described process of approximating in 4D works wonderfully; however, there are serious problems when it comes to rendering the same surface using shading.

The usually accepted shading method can be broken down into the following steps.

i)   The surface to be rendered is approximated by planar polygons.

ii)   At each vertex of the polygons a number of values have to be calculated to evaluate the shading equations. These data depend on the reflection type (diffuse, specular) and the light source types (ambient, positional, directional, spot) and include typically such data as normal vectors (the shading equation are presented in the appendix of the PHIGS PLUS document [ISO89a] and are also explained in more detailed and in a much more understandable way in for example [Pöps89]).

iii)   The polygons are rendered individually by interpolating the values determined by step *ii)* for the interior pixels. Whether this interpolation involves the colour values only (Gouraud shading) or the vector values as well (Phong shading) is usually a settable parameter.

The source of the problem lies in step *ii)*. Indeed, the polynomial approximation can be done without problems in 4D and the interpolation step itself does not depend on the geometrical features. On the other hand, step *ii)* involves a number of geometrical data which are all Euclidean in nature (see also figure 8.2 as well as the shading equations described in [ISO89a] or [Pöps89]): normals, vectors and distances between the points on the surface and the light source, direction of viewing etc. All these data are distorted by a projective transformation, that is for example the image

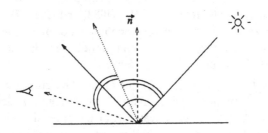

Figure 8.2.

of $\vec{n}$ will not be the normal on the image of the surface any more; that is it is not mathematically equivalent to approximate the surface and the corresponding vectors first and then transform them or the other way round. In other words, if step *ii)* is performed in 4D, after the linear part of the transformation, the visual effect of the shading calculations will be different; the only way of following steps exactly *i)-iii)* is to perform them *before* the transformation. This is quite a disturbing fact; indeed, the B–Spline formulation has the very attractive feature of describing a complicated shape with a relatively low number of points (eg the description of a full torus in space requires 64 points in space with weights, see [Pieg87]) whereas the correct visualisation of the same surface would require much more. It makes therefore quite a difference whether the full approximation has to be made before the transformation (leading to a huge number of matrix–vector multiplications in practice) or afterwards.

There is no known solution to this problem. In a recent paper, Abi–Ezzi and Wozny ([Abie90], see also [Abie89]) give a very pragmatic solution by "factoring" a projective transformation; this means the transformation is described as a concatenation of a fully orthogonal and a projective transformation, where this latter one has a relatively "sparse" matrix (that is it can be proven that a number of elements will be zero). The approximation and the shading equations are evaluated after the orthogonal transformation but before the projective one. This solution works well and leads to speed improvements but is clearly not a full one.

A possible approach for a more elaborate solution would be to revise the shading model itself. It should not be forgotten that all shading equations (as well as the interpolations of step *iii)*) are only models and approximations of physical phenomena; an alternative method could eventually be found which would lead to a projective invariant formulation. This alternative should be, of course, no worse in its visual effect than the old one. What such a new model could look like is however still an open question.

References

[Abie89] S.S. Abi-Ezzi, *The Graphical Processing of B–Splines in a Highly Dynamic Environment*, PhD Thesis, Rensselaer Polytechnic Institute, Troy, New York (1989).

[Abie90] S.S. Abi-Ezzi and M.J. Wozny, "Factoring a Homogeneous Transformation for a more Efficient Graphics Pipeline", in *Eurographics'90 Conference Proceedings*, eds. C.E. Vandoni and C.E. Vandoni, North Holland, Amsterdam (1990). Also in *Computer Graphics Forum*, **9** (1990).

[Anjy90] K. Anjyo, "Mathematical Models for Semi–globalized Spectral Syntheis", in *Eurographics'90 Conference Proceedings*, eds. C.E. Vandoni and C.E. Vandoni, North Holland, Amsterdam (1990)

[Arde87] "Doré Programmer's Guide", Release 1.0, Ardent Computer Corporation, Sunnyvale, CA. (1987).

[Arno90] D.B. Arnold and D.A. Duce, *ISO Standards for Computer Graphics: The First Generation*, Butterworths, London (1990).

[Arok89] A. Arokiasamy, "Homogeneous Coordinates and the Principle of Duality in Two Dimensional Clipping", *Computers and Graphics*, **13** (1989).

[Bart87] R.H. Bartels, J.C. Beatty and B.A. Barsky, *An Introduction to Splines for Use in Computer Graphics & Geometric Modelling*, Morgan Kaufmann Publishers, Inc., Los Altos, CA. (1987).

[Berm61] G. Berman, "The Wedge Product", *American Mathematical Monthly*, **68** (1961).

[Bez83] H.E. Bez, "Homogeneous Coordinates for Computer Graphics", *Computer Aided Design*, **15** (1983).

[Blin78] J.F. Blinn and M.E. Newell, "Clipping Using Homogeneous Coordinates", *Computer Graphics*, **12** (1978).

[Clif88] W. Clifford, J.I. McConnell and J. Saltz, "The Development of PEX", in *Eurographics'88 Conference Proceedings*, eds. D.A. Duce and P. Jancène, North-Holland, Amsterdam (1988).

[GSPC77] ACM Siggraph GSPC, "Status Report of the Graphics Standards Planning Committee", *Computer Graphics*, **11** (1977).

[GSPC79] ACM Siggraph GSPC, "Status Report of the Graphics Standards Planning Committee", *Computer Graphics*, **13** (1979).

[Coxe49] H.S.M. Coxeter, *The Real Projective Plane*, McGraw-Hill, New York - Toronto - London (1949).

[Coxe74] H.S.M. Coxeter, *Projective Geometry*, University of Toronto Press, Toronto (1974).

[Crem60] L. Cremona, *Elements of Projective Geometry*, Dover Publications, Inc., New York (1960).

142

[Düre66] A. Dürer, *Underweysung der Messung mit dem Zirckel un Richtscheyt in Linien Ebnen un Ganzen Corporen*, (Facsimile Reprint from the original of 1525), Josef Stocker/Schmind, Nürnberg (1966).

[Eukl75] Euklid, *Die Elemente, Buch I-XIII*, Akademische Verlagsgesellschaft, Leipzig (1975).

[Fari88] G. Farin, *Curves and Surfaces for Computer Aided Geometric Design*, Computer Science and Scientific Computing, Academic Press, San Diego - London (1988).

[Fäus71] A. Fäustle, *Technisches Zeichnen*, Don Bosco Verlag, München (1971).

[Faux79] I.D. Faux and M.J. Pratt, *Computational Geometry for Design and Manifacture*, Ellis Horwood, New York - Chichester - Brisbane - Toronto (1979).

[Fisc85] G. Fischer, *Analytische Geometrie*, Vieweg Studium, Grundkurs Mathematik, Vieweg and Sohn, Braunschweig (1985).

[Fole84] J.D. Foley and A. van Dam, *Fundamentals of Computer Graphics*, Addison–Wesley, Reading, MA. (1984).

[Fole90] J.D. Foley, A. van Dam, S.K. Feiner and J.F. Hughes, *Computer Graphics: Principles and Practice*, Addison–Wesley, Reading, MA. (1990).

[Ghar85] N. Gharachorloo and Ch. Pottle, "SUPER BUFFER: A Systolic VLSI Graphics Engine for Real Time Raster Image Generation", in *Proceedings of Chapel Hill Conference on VLSI*, Computer Science Press, Rockville, MD. (1985). Also in *Tutorial: Computer Graphics Hardware, Image Generation and Display*, eds. H.K. Reghbati and A.Y.C. Lee, IEEE Computer Society Press, Washington D.C. (1988).

[Görö88] J. Görög, G. Krammer and A. Vincze, "IXPHIGS: A Portable Implementation of the International PHIGS Standard", in *Eurographics'88 Conference Proceedings*, eds. D.A Duce and P. Jancène, North Holland, Amsterdam (1988).

[Hage87] P.J.W. ten Hagen, A.A.M. Kuijk and C.G. Trienekens, "Display Architecture for VLSI-based Graphics Workstations", in *Advances in Computer Graphics Hardware I*, ed. W. Straßer, EurographicSeminar Series, Springer Verlag, Berlin - Heidelberg - New York - Tokyo (1987).

[Hage90] P.J.W. ten Hagen, I. Herman and J.R.G. de Vries, "A Dataflow Graphics Workstation", *Computer and Graphics*, 14, (1990). Also in *Reports of the Centre for Mathematics and Computer Sciences*, Report No. CS-R8910, Amsterdam (1989).

[Hajó60] Gy. Hajós, *Bevezetés a Geometriába*, Tankönyvkiadó, Budapest (1960).

[Heck86] P.S. Heckbert, "A Survey of Texture Mapping", *IEEE Computer Graphics & Applications*, 6 (1986).

[Heck89] P.S. Heckbert, "Fundamentals of Texture Mapping and Image Warping", *Reports of the Computer Science Divisions of the University of California,*

Report No. UCB/CSD 89/156, Berkeley, CA. (1989).

[Herm87] I. Herman and J. Reviczky, "A Means to Improve the GKS-3D/PHIGS Output Pipeline Implementation", in *Eurographics'87 Conference Proceedings*, ed. G. Maréchal, North Holland, Amsterdam, (1987). Also in *Computers and Graphics*, **12** (1988).

[Herm88] I. Herman and J. Reviczky, "Some Remarks on the Modelling Clip Problem", *Computer Graphics Forum*, **7** (1988).

[Herm88a] I. Herman, T. Tolnay-Knefély, J. Reviczky and F.L. Westhoff, "Three Dimensional Graphics Standards and CGI", *Computers and Graphics*", **12**, (1988).

[Herm89] I. Herman, "2.5D Graphics Systems", in *Eurographics'89 Conference Proceedings*, eds. W. Hansmann, F.R.A. Hopgood and W. Straßer, North-Holland, Amsterdam (1989). Also in ISO document ISO/IEC JTC 1/SC 24 API-13 (1989). Also in *Reports of the Centre for Mathematics and Computer Sciences*, Report No. CS-R8921, Amsterdam (1989).

[Herm89a] I. Herman, "On The Projective Invariant Representation of Conics in Computer Graphics", *Computer Graphics Forum*, **8** (1989). Also in *Reports of the Centre for Mathematics and Computer Sciences*, Report No. CS-R8938, Amsterdam (1989).

[Herm90] I. Herman, "The Use of Projective Geometry in Computer Graphics", PhD Thesis, University of Leiden (1990).

[Herm91] I. Herman, "Projective Geometry and Computer Graphics", in *Advances in Computer Graphics IV*, eds. W.T. Hewitt, M. Grave and M. Roch, EurographicSeminar Series, Springer Verlag, Berlin - Heidelberg - New York - Tokyo (1991).

[Hest84] D. Hestenes and G. Sobczyk, *Clifford Algebra to Geometric Calculus: A Unified Language for Mathematics and Physics*, D. Reidel, Dordrecht, (1984).

[Heyt63] A. Heyting, *Projective Meetkunde*, P. Noordhoff N.V., Groningen (1963).

[Hopg83] F.R.A. Hopgood, D.A. Duce, J.R. Gallop and D.C. Sutcliffe, *Introduction to the Graphical Kernel System GKS*, Academic Press, London - New York (1983).

[Hopg91] F.R.A. Hopgood and D.A. Duce, *A primer for PHIGS*, John Wiley & Sons (1991).

[Howa87] T.L.J. Howard, "A Shareable Centralised Database of KRT^3 - A Hierarchical Graphics System Based on PHIGS", in *Eurographics'87 Conference Proceedings*, ed. G. Maréchal, North Holland, Amsterdam, (1987). Also in *Computers and Graphics*, **12** (1988).

[Howa91] T.L.J. Howard, W.T. Hewitt, R.J. Hubbold and K.M. Wyrwas, *A Practical Introduction to PHIGS and PHIGS PLUS*, Addison-Wesley, Workingham - Reading (1991).

[Hübl90] J. Hübl and I. Herman, "Modelling Clip: Some More Results", *Computer Graphics Forum*, **9** (1990). Also in *Reports of the Centre for Mathematics and Computer Sciences*, Report No. CS-R9008, Amsterdam (1990).

[Hubb87] R.J. Hubbold, "3D Graphics Standards: A Critical Appraisal", in *Ausgraph'87 Conference Proceedings*, Australian Computer Graphics Association (1987).

[Hubb90] R. Hubbold and T. Hewitt "GKS-3D and PHIGS: Theory and Practice", in *Advances in Computer Graphics IV*, eds. W.T. Hewitt, M. Grave and M. Roch, EurographicSeminar Series, Springer Verlag, Berlin - Heidelberg - New York - Tokyo (1990).

[ISO85] "Information processing systems - Computer graphics - Graphical Kernel System (GKS), functional description", ISO 7942 (1985).

[ISO88] "Information processing systems - Computer graphics - Graphical Kernel System for Three Dimensions (GKS-3D), functional description", ISO 8805 (1988).

[ISO88a] "Information processing systems - Computer graphics - Interfacing techniques for dialogues with graphical devices (CGI), functional description", ISO DP 9636/1-6 (1988).

[ISO88b] "PEX Protocol Specification", ISO IEC JTC 1/SC 24/WG2/N2 (1988).

[ISO89] "Information processing systems - Computer graphics, Programmer's Hierarchical Interactive Graphics System (PHIGS) - Part 1, functional description", ISO/IEC 9592-1 (1989).

[ISO89a] "Information processing systems - Computer graphics, Programmer's Hierarchical Interactive Graphics System (PHIGS) - Part 4, Plus Lumière und Surfaces (PHIGS PLUS)", ISO/IEC 9592-4 (1990).

[Kell63] O.H. Keller, *Analytische Geometrie und Lineare Algebra*, Deutscher Verlag der Wissenschaften, Berlin (1963).

[Keré66] B. Kerékjártó, *Les Fondements de la Géométrie, Vol. II, Géométrie Projective*, Akadémiai Kiadó, Budapest (1966).

[Klem89] K. Klement, "Allgemeine Rotationsflächen und Deren Darstellung als Rationale Flächen", *CAD und Computergraphik*, **12** (1989).

[Kram89] G. Krammer, "Notes on the Mathematics of the PHIGS Output Pipeline", *Computer Graphics Forum*, **8** (1989).

[Lanc70] C. Lanczos, *Space Through the Ages*, Academic Press, Inc., London (1970).

[Magn86] N. Magnenat-Thalmann, D. Thalmann, "Introduction à l'Informatique Graphique", in *Advances in Computer Graphics I*, eds. G. Enderle, M. Grave and F. Lillehagen, EurographicSeminar Series, Springer Verlag, Berlin - Heidelberg - New York - Tokyo (1986).

[Mudu86] S.P. Mudur, "Mathematical Elements for Computer Graphics", in *Advances in Computer Graphics I*, eds. G. Enderle, M. Grave and F. Lillehagen, EurographicSeminar Series, Springer Verlag, Berlin - Heidelberg - New York - Tokyo (1986).

[Márt86] S. Márton, *Nincs királyi út (Matematikatörténet)*, Gondolat, Budapest (1986).

[Newm79] W.M. Newmann and R.F. Sproull, *Principles of Interactive Computer Graphics*, McGraw-Hill, New York - Toronto - London (1979).

[OBar89] R. M. O'Bara and S. S. Abi-Ezzi, "An Analysis of Modeling Clip", in *Eurographics'89 Conference Proceedings*, eds. W. Hansmann, F.R.A. Hopgood and W. Straßer, North-Holland, Amsterdam (1989).

[Penn86] M.A. Penna and R.R. Patterson, *Projective Geometry and Its Application to Computer Graphics*, Prentice-Hall, New Jersey (1986).

[Pieg87] L. Piegl and W. Tiller, "Curve and Surface Constructions Using Rational B-Splines", *Computer Aided Design*, **19** (1987).

[Pixa88] "The RenderMan Interface", Version 3.0, PIXAR, San Rafael, CA. (1988).

[Pun90] Th. Pun and E. Blake, "Relationships Between Image Synthesis and Analysis: Toward Unification?", Report of the Eurographics Working Group on Relationships Between Image Synthesis and Analysis, *Computer Graphics Forum*, **9** (1990).

[Pöps89] J. Pöpsel and Ch. Hornung, "HighlightShading, Lighting and Shading in a PHIGS+/PEX Environmnent", in *Eurographics'89 Conference Proceedings*, eds. W. Hansmann, F.R.A. Hopgood and W. Straßer, North-Holland, Amsterdam (1989). Also in *Computers and Graphics*, **14** (1990).

[Reis81] R.F. Reisenfeld, "Homogeneous Coordinates and Projective Planes in Computer", Graphics *IEEE Computer Graphics & Application*, **1** (1981).

[Rose63] R.A. Rosenbaum, *Introduction to Projective Geometry and Modern Algebra*, Addison-Wesley, Reading, MA. (1963).

[Suth74] I.E. Sutherland and G.W. Hodgman, "Reentrant Polygon Clipping", *Comm. of the ACM*, **17** (1974).

[Sabi77] M.A. Sabin, "The Use of Piecewise Forms for the Numerical Representation of Shape", PhD Thesis, Reports of the Computer and Automation Institute of the Hungarian Academy of Sciences, 60/1977, Budapest (1977).

[Salm87] M. Salmon and R. Slater, *Computer Graphics, Systems and Concepts*, Addison-Wesley, Reading, MA. (1987).

[Seid90] H.-P. Seidel, "Quaternionen in Computergraphik und Robotik", in *Geometrische Verfahren der Graphischen Datenverarbeitung*, ZGDV-Reihe Beiträge zur Graphischen Datenverarbeitung, Springer Verlag, Berlin - Heidelberg - New York - Tokyo (1990).

[Sing86] K. Singleton, "An Implementation of the GKS-3D/PHIGS Viewing Pipe-line", in *Eurographics'86 Conference Proceedings*, ed. A.A.G. Requicha, North-Holland, Amsterdam (1986). Also in *GKS - Theory and Practice*, eds. P.R. Bono and I. Herman, EurographicSeminar Series, Springer Verlag, Berlin - Heidelberg - New York - Tokyo (1987).

[Stol89] J. Stolfi, "Primitives for Computational Geometry", Digital Systems Research Center Report 36, Palo Alto, CA. (1989).

[Stru53] D.J. Struik, *Lectures on Analytic and Projective Geometry*, Addison-Wesley, Reading, MA. (1953).

[Till83] W. Tiller, "Rational B-Splines for Curve and Surface Representation", *IEEE Computers Graphics & Applications*, 3 (1983).

[Vára84] T. Várady, "Basic Equations and Simple Geometric Properties of Double-quadratic Curves and Surfaces", *CAD Group Document 117*, Cambridge University Engineering Department (1984).

[Vára85] T. Várady, "Integration of Free-form Surfaces Into a Volumetric Modeller", PhD Thesis, Reports of the Computer and Automation Institute of the Hungarian Academy of Sciences, 171/1985, Budapest (1985).

[Watt89] A. Watt, *Three Dimensional Computer Graphics*, Addison Wesley, Wokingham (1989).

[Wood71] P.A. Woodsford, "The Design and Implementation of the GINO 3D Graphics Software Package", Software Practice and Experience, 1 (1971).

[Zach89] M. Zachrisen, "Yet Another Remark on the Modelling Clip Problem", *Computer Graphics Forum*, 8 (1989).

Lecture Notes in Computer Science

For information about Vols. 1–481
please contact your bookseller or Springer-Verlag

1991.

Vol. 524: G. Rozenberg (Ed.), Advances in Petri Nets 1991. VIII, 572 pages. 1991.

Vol. 525: O. Günther, H.-J. Schek (Eds.), Advances in Spatial Databases. Proceedings, 1991. XI, 471 pages. 1991.

Vol. 526: T. Ito, A. R. Meyer (Eds.), Theoretical Aspects of Computer Software. Proceedings, 1991. X, 772 pages. 1991.

Vol. 527: J.C.M. Baeten, J. F. Groote (Eds.), CONCUR '91. Proceedings, 1991. VIII, 541 pages. 1991.

Vol. 528: J. Maluszynski, M. Wirsing (Eds.), Programming Language Implementation and Logic Programming. Proceedings, 1991. XI, 433 pages. 1991.

Vol. 529: L. Budach (Ed.), Fundamentals of Computation Theory. Proceedings, 1991. XII, 426 pages. 1991.

Vol. 530: D. H. Pitt, P.-L. Curien, S. Abramsky, A. M. Pitts, A. Poigné, D. E. Rydeheard (Eds.), Category Theory and Computer Science. Proceedings, 1991. VII, 301 pages. 1991.

Vol. 531: E. M. Clarke, R. P. Kurshan (Eds.), Computer-Aided Verification. Proceedings, 1990. XIII, 372 pages. 1991.

Vol. 532: H. Ehrig, H.-J. Kreowski, G. Rozenberg (Eds.), Graph Grammars and Their Application to Computer Science. Proceedings, 1990. X, 703 pages. 1991.

Vol. 533: E. Börger, H. Kleine Büning, M. M. Richter, W. Schönfeld (Eds.), Computer Science Logic. Proceedings, 1990. VIII, 399 pages. 1991.

Vol. 534: H. Ehrig, K. P. Jantke, F. Orejas, H. Reichel (Eds.), Recent Trends in Data Type Specification. Proceedings, 1990. VIII, 379 pages. 1991.

Vol. 535: P. Jorrand, J. Kelemen (Eds.), Fundamentals of Artificial Intelligence Research. Proceedings, 1991. VIII, 255 pages. 1991. (Subseries LNAI).

Vol. 536: J. E. Tomayko, Software Engineering Education. Proceedings, 1991. VIII, 296 pages. 1991.

Vol. 537: A. J. Menezes, S. A. Vanstone (Eds.), Advances in Cryptology – CRYPTO '90. Proceedings. XIII, 644 pages. 1991.

Vol. 538: M. Kojima, N. Megiddo, T. Noma, A. Yoshise, A Unified Approach to Interior Point Algorithms for Linear Complementarity Problems. VIII, 108 pages. 1991.

Vol. 539: H. F. Mattson, T. Mora, T. R. N. Rao (Eds.), Applied Algebra, Algebraic Algorithms and Error-Correcting Codes. Proceedings, 1991. XI, 489 pages. 1991.

Vol. 540: A. Prieto (Ed.), Artificial Neural Networks. Proceedings, 1991. XIII, 476 pages. 1991.

Vol. 541: P. Barahona, L. Moniz Pereira, A. Porto (Eds.), EPIA '91. Proceedings, 1991. VIII, 292 pages. 1991. (Subseries LNAI).

Vol. 543: J. Dix, K. P. Jantke, P. H. Schmitt (Eds.), Nonmonotonic and Inductive Logic. Proceedings, 1990. X, 243 pages. 1991. (Subseries LNAI).

Vol. 544: M. Broy, M. Wirsing (Eds.), Methods of Programming. XII, 268 pages. 1991.

Vol. 545: H. Alblas, B. Melichar (Eds.), Attribute Grammars, Applications and Systems. Proceedings, 1991. IX, 513 pages. 1991.

Vol. 547: D. W. Davies (Ed.), Advances in Cryptology – EUROCRYPT '91. Proceedings, 1991. XII, 556 pages. 1991.

Vol. 548: R. Kruse, P. Siegel (Eds.), Symbolic and Quantitative Approaches to Uncertainty. Proceedings, 1991. XI, 362 pages. 1991.

Vol. 550: A. van Lamsweerde, A. Fugetta (Eds.), ESEC '91. Proceedings, 1991. XII, 515 pages. 1991.

Vol. 551:S. Prehn, W. J. Toetenel (Eds.), VDM '91. Formal Software Development Methods. Volume 1. Proceedings, 1991. XIII, 699 pages. 1991.

Vol. 552: S. Prehn, W. J. Toetenel (Eds.), VDM '91. Formal Software Development Methods. Volume 2. Proceedings, 1991. XIV, 430 pages. 1991.

Vol. 553: H. Bieri, H. Noltemeier (Eds.), Computational Geometry - Methods, Algorithms and Applications '91. Proceedings, 1991. VIII, 320 pages. 1991.

Vol. 554: G. Grahne, The Problem of Incomplete Information in Relational Databases. VIII, 156 pages. 1991.

Vol. 555: H. Maurer (Ed.), New Results and New Trends in Computer Science. Proceedings, 1991. VIII, 403 pages. 1991.

Vol. 556: J.-M. Jacquet, Conclog: A Methodological Approach to Concurrent Logic Programming. XII, 781 pages. 1991.

Vol. 557: W. L. Hsu, R. C. T. Lee (Eds.), ISA '91 Algorithms. Proceedings, 1991. X, 396 pages. 1991.

Vol. 558: J. Hooman, Specification and Compositional Verification of Real-Time Systems. VIII, 235 pages. 1991.

Vol. 559: G. Butler, Fundamental Algorithms for Permutation Groups. XII, 238 pages. 1991.

Vol. 560: S. Biswas, K. V. Nori (Eds.), Foundations of Software Technology and Theoretical Computer Science. Proceedings, 1991. X, 420 pages. 1991.

Vol. 561: C. Ding, G. Xiao, W. Shan, The Stability Theory of Stream Ciphers. IX, 187 pages. 1991.

Vol. 562: R. Breu, Algebraic Specification Techniques in Object Oriented Programming Environments. XI, 228 pages. 1991.

Vol. 563: A. Karshmer, J. Nehmer (Eds.), Operating Systems of the 90s and Beyond. Proceedings, 1991. X, 285 pages. 1991.

Vol. 564: I. Herman, The Use of Projective Geometry in Computer Graphics. VIII, 146 pages. 1992.